AF694765

Non-linearities in Passive RFID Systems

FOCUS SERIES

Non-linearities in Passive RFID Systems

Third Harmonic Concept and Applications

Gianfranco Andía
Yvan Duroc
Smail Tedjini

ISTE

WILEY

First published 2018 in Great Britain and the United States by ISTE Ltd and John Wiley & Sons, Inc.

Apart from any fair dealing for the purposes of research or private study, or criticism or review, as permitted under the Copyright, Designs and Patents Act 1988, this publication may only be reproduced, stored or transmitted, in any form or by any means, with the prior permission in writing of the publishers, or in the case of reprographic reproduction in accordance with the terms and licenses issued by the CLA. Enquiries concerning reproduction outside these terms should be sent to the publishers at the undermentioned address:

ISTE Ltd
27-37 St George's Road
London SW19 4EU
UK

www.iste.co.uk

John Wiley & Sons, Inc.
111 River Street
Hoboken, NJ 07030
USA

www.wiley.com

© ISTE Ltd 2018
The rights of Gianfranco Andía, Yvan Duroc and Smail Tedjini to be identified as the authors of this work have been asserted by them in accordance with the Copyright, Designs and Patents Act 1988.

Library of Congress Control Number: 2017958908

British Library Cataloguing-in-Publication Data
A CIP record for this book is available from the British Library
ISSN 2051-2481 (Print)
ISSN 2051-249X (Online)
ISBN 978-1-78630-226-7

Contents

Chapter 5. [illegible]

Acknowledgments

Le Temps de Lecteurs

Dans une taverne de Valence
Où se retrouvaient ***tous les chercheurs***
Nous discutions de radiofréquence
Des étiquettes, ainsi que des lecteurs....

C'était le temps des fleurs
On ignorait la peur
Le mot "docteur" *avait un goût de miel*
Ton bras prenait mon bras,
Ta voix suivait ma voix,
On était jeune
Et l'on croyait au ciel...

Et puis sont venus les jours de brume
Avec des bruits étranges et des pleurs
Combien j'ai passé de nuit sans lune
A chercher l'étiquette avec le lecteur...

Et ce soir je suis devant la porte
De la taverne ici même à Valence
Et la chanson que la nuit m'apporte
Il y a beaucoup de radiofréquences!

Tout comme au temps des fleurs
Où l'on vivait sans peur
Où chaque jour avait un goût de miel
Ton bras prenait mon bras,
Ta voix suivait ma voix ...
Aujourd'hui nous
Croyons encore au ciel!

Pavel Nikitin, Valence, November 2014

The lyrics of this song are credited to Pavel Nikitin, after modifying the lyrics of "Le temps des fleurs" by Dalida.

Pavel was in Valence, France in November 2014 as jury member in the PhD dissertation of Gianfranco and also as speaker in the workshop "Hot topics in RFID/NFC". These lyrics represent the great memories the authors of this book have about Pavel's visit, in both scientific and emotional aspects. They would like to celebrate the sincere and fruitful friendship between Yvan, Smail and Gianfranco, and await the possibility of many further joint projects!

About the song

The original version of "Le temps des fleurs" is a Russian romance song "Dorogoi dlinnoyu" ("Дорогой длинною", literally "By the long road"), composed by Boris Fomin (1900–1948) with words by the poet Konstantin Podrevsky. Gene Raskin then adapted it into English and named it "Those Were the Days", after which it was adapted to French and sung by Dalida.

Introduction

Since its early commercial application as an anti-theft system in supermarkets, the radio-frequency identification (RFID) technology has become a ubiquitous technology not only in industrial or agricultural areas but also in our daily lives, for instance our car key or access badge. Being able to identify a person, an animal, an object and even hundreds of objects simultaneously has opened up many other applications such as indoor localization and tracking, and further applications involving integration of sensors with RFID. As with the most popular wireless technologies, such as Wi-Fi, Bluetooth or NFC, passive ultra-high-frequency (UHF) RFID has rapidly grown thanks to the standards but also the promotion by industrial organizations such as RAIN RFID Alliance. Nowadays, passive UHF RFID is reliable and technologically and economically lucrative. Although based on relatively simple concepts compared to other wireless technologies using multiple antennas or complex modulations, RFID offers many degrees of liberty that allow the opening of new perspectives and very promising applications. With the idea of preserving devices as simply as possible and enabling the identification of something or somebody, the latest technological achievements in terms of low-power consumption microelectronics along with the exploitation of signal properties not yet used today allow important enhancements in the performance of the RFID system. These improvements include longer read ranges, accurate inventories with high density of tags, data rate and security, which are used in increasingly harsh environments. In parallel, continuous innovation has launched RFID technology as the Internet enabler dedicated to the last meters of tagged things. We can say that the simple backscattering modulation principle has set the path for an open road; the RFID revolution has begun.

In this context, this book focuses on the exploitation of the harmonic signals that are generated by RFID chips in compliance with the passive UHF RFID standard. Even if the topic is all based on RFID, the concepts introduced may also find applications in other fields as long as harmonic signals are present. Chapter 1 introduces RFID technology and its history. It provides the details about current perspectives and expectations in this field. Chapter 2 summarizes the technical fundamentals of the RFID technology and highlights the non-linear circuits present in passive RFID chips and the source of so far "wasted" signals, i.e. harmonic signals existing but not yet used. Chapter 3 presents recent and innovative experimental methods to characterize RFID tags and chips, which provide information about the production of these harmonic signals during a standard RFID communication. Chapter 4 presents theoretical approaches that model and analyze the observed non-linear behavior source of harmonic generation. The study can be used to minimize the "naturally" harmonic signals transmitted by commercial passive UHF RFID tags. On the contrary, in Chapter 5, applications and associated prototypes demonstrate that these unwanted signals can become either an information source with a harmonic tag or an energy source.

1

History of Radio-frequency Identification: from Birth to Advanced Applications

Whether we are aware of it or not, RFID (radio-frequency identification) technology is an integral part of our everyday life. The RFID concept was anticipated a century ago, and nowadays it continues to advance as technology progresses and rapidly evolves to have more impact on our modern societies. The objective of this chapter is to give an account of the development of RFID from its birth to the present day.

1.1. Early facts about the genesis of RFID

The US patent 17744036, first filled in France in 1924 [BRA 30], was certainly one of the very first, if not the first, documents that described the concept of RFID technology. The patent describes the novelty as follows: "*The primary station comprises a source of oscillations and an emitting circuit; the secondary has a receiving circuit with no local energy source, and in which is inserted a manipulating or modulating device (telephonic or telegraphic apparatus)*". This is no more than the principle of the RFID technology that exploits the reflected waves in order to ensure the communication between two devices: one of them is fully passive (the tag or the transponder) as it does not embed any energy source; however, it converts part of the signal sent by the other device (the reader or the interrogator) to modulate the refection towards the reader.

Two decades before this patent publication, H. Hertz noted that electromagnetic waves were reflected by metallic objects, even if he did not apply this insight in a practical engineering sense as done by others. G. Marconi was the first to show that these waves could be used for communication and also noted that radio waves were being reflected back to the transmitter by objects in radio beacon experiments that he conducted on March 3, 1899 on Salisbury Plain. Radar applications had just been born and their evolution from then will not cease and will be one of the powerful engines of RF (radio frequency) development and wireless applications. In 1904, C. Hülsmeyer [HUE 06] gave public demonstrations in Germany and the Netherlands of the use of radio echoes to detect ships so that collisions could be avoided. Moreover, in 1915, R. Watson-Watt joined the Meteorological Office as a meteorologist at Aldershot in Hampshire; over the next few years, he studied atmospheric phenomena and developed the use of radio signals generated by lightning strikes to map out the position of thunderstorms. Later, in the 1930s, the emergence of the radar system and its very rapid adoption by the military during World War II pushed engineers to explore the high-frequency domain and to master various techniques for signal generation, propagation and detection as well as broadcast. In 1935, H. E. Hollmann wrote one of the first comprehensive books on microwaves and applications. The book was titled “Physics and Technique of Ultrashort Waves”. This book was used throughout many parts of the world, which favored the development of radar technology in many countries.

1.2. Birth of RFID

Although there is no logical progression to RFID technology and often key advances are not recognized until later, developments in radar and broadcasting have evidently profited RFID. Among the variety of historical facts that paved the way for the emergence of modern RFID, two major events can be considered as the cornerstones. The first is a theoretical paper that very precisely addressed the physics behind the operating principle of RFID. This is the landmark paper published by H. Stockman, in the Proceedings of the IRE, October 1948. The paper was entitled “Communication by Means of Reflected Power”. Stockman stated in his paper that “Evidently, considerable research and development work has to be done before the remaining basic problems in reflected-power communication are solved, and before the field of useful applications is explored”. The second fact is more practical as it regards the design and application of a device that is probably the first tag ever realized. Such a device was developed by L. Theremin, which was used as a passive wireless microphone for spying [NIK 12]. A schematic diagram of this device, known in the literature as "The Thing", is shown in Figure 1.1. The Thing is composed of a metallic box that acts as an electromagnetic cavity with a specific resonance frequency coupled to a monopole antenna. To transform the device into a microphone, one wall of the cavity is replaced by a metallic vibrating membrane that is sensitive to sound waves. When the device is illuminated by an electromagnetic continuous wave at the resonance frequency of the system, a reflected signal (backscatter signal) at the same frequency is generated and can be detected with a dedicated system. If there is no sound, the amplitude of the backscatter signal will be constant. However, if some sound waves reach the vibrating membrane, the backscatter signal will be modulated by the sound and therefore detectable thanks to the simple envelope detector. As we can see, the system is so simple and its principle of operation is straightforward! The operation frequency was approximately 330 MHz. The beauty of the system lies in the fact that it had really been used and had worked for several years. Indeed, the device was embedded in a carved wooden plaque of the US Great Seal, and on August 4, 1945, Soviet school children presented it to the US ambassador as a "gesture of friendship". It hung in the ambassador's Moscow residential office and served to spy on the conversations of the ambassador for years. The existence of the bug was accidentally discovered by a British radio operator who overheard American conversations on an open radio channel as the Russians were beaming radio waves at the ambassador's office.

A third fact is often cited as part of the history of RFID. This is the IFF (Identify Friend or Foe) that was developed and exploited by the British during World War II [BRO 00]. The operating frequency of the system was in the range of 20–30 MHz. First, the system was passive and used wires across the wings of the airplane with open/closed operations controlled by motor relay to modulate the backscatter signal. Later, the system evolved into an active solution with amplifier and modern electronics. Nowadays, the IFF enables military and civilian air traffic control interrogation systems to identify aircraft and to determine their bearing and range from the interrogator.

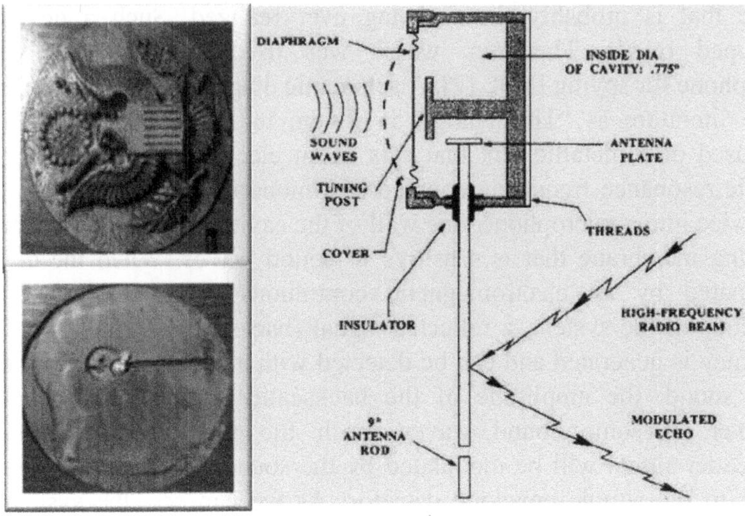

Figure 1.1. *Depiction of the device designed by L. Theremin and used as a wireless passive microphone. Left: the device as it was embedded in a carved wooden plaque of the US Great Seal. Right: the composition of "The Thing"*

Interestingly, the 1950s was the decade that explored the RFID techniques that emerged from the development of radar and broadcasting that took place in the two previous decades. Apart from the IFF, developments during the 1950s included works such as that of D.B. Harris dedicated to "Radio transmission systems with modulable passive responder" [HAR 60].

1.3. Early modern RFID

During the 1960s, thanks to the advances in electronic technology, it was possible to develop the first commercial tag, i.e. the electronic article surveillance (EAS) tag. The first tags deployed in the 1960s were hard, round and plastic. These tags are based on the resonance properties of a simple LC tank circuit. The main application of EAS is to detect the presence or absence of the LC circuit thanks to its resonance when coupled to the sweep generator used as a reader. A schematic description of the EAS system is given in Figure 1.2. The LC circuit that represents the EAS tag can be turned off if its capacitor is deactivated. Such a deactivation occurs when the reader sends an interrogation signal with a very large amplitude at the resonance frequency of the system.

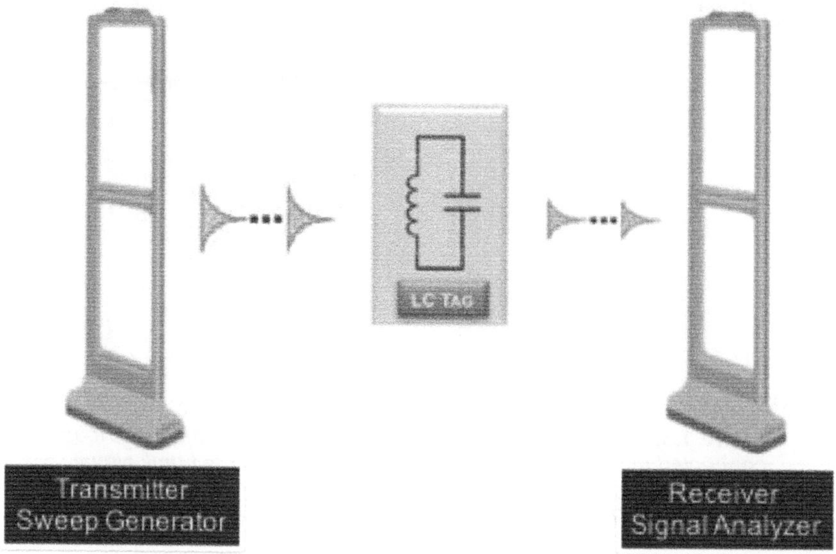

Figure 1.2. *Description of the EAS system. The tag is primarily an LC circuit whose presence is detected by a reader which is primarily a sweep generator*

Since their introduction, EAS tags have undergone remarkable evolution thanks to the huge advances in electronic and microelectronic technologies. Although EAS tags were first developed for apparel retailers to avoid thefts, they have since been applied to a wider range of goods, such as books in libraries, video and music products, etc. Companies such as Sensormatic and Checkpoint were founded in the late 1960s which, with others, developed EAS equipment to counter theft. These types of systems are often known as “1-bit” tags, because only the presence or absence of a tag could be detected. The first EAS tags introduced in the 1960s were rigid and quite big. In the 1980s, the tags became smaller and soft so that they could be attached with adhesive backs, as well as deactivated at the sale store. Later, in the 1990s, the tags could be sewn into or heat-sealed onto items of clothing at the manufacture point [DIL 15]. Therefore, EAS tags have become the most commonly used article surveillance and anti-theft technique. They are boosted by more and more cheap integration technologies, and can be made inexpensively and serve as an effective anti-theft system. Examples of EAS tags are shown in Figure 1.3.

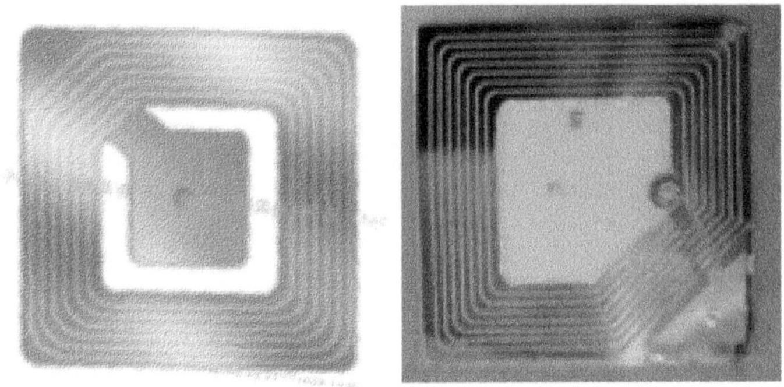

Figure 1.3. *Examples of EAS tags. We clearly see the magnetic loop connected to the chip capacitor to form the LC resonant tank circuit. Typical operation frequencies are in the range of 1.75 MHz to 9.5 MHz band. The standard frequency for retail use is 8.2 MHz*

As far as the physics behind RFID is considered, the 1960s was the prelude to several very consistent theoretical publications and practical demonstrations and inventions [LAN 05]. Among the major contributions are those of R. F. Harrington who studied the electromagnetic theory related to RFID, which was published in his papers “Field measurements using active scatterers” and “Theory of loaded scatterers” in 1963–1964. Regarding the inventions, R. Richardson developed “Remotely activated radio frequency powered devices” in 1963, O. Rittenback introduced “Communication by radar beams” in 1969, J. H. Vogelman demonstrated “Passive data transmission techniques utilizing radar beams” in 1968 and J. P. Vinding performed “Interrogator–responder identification system” in 1967. These theoretical contributions and practical advances in the developments of radar and broadcast profited the birth of modern RFID and its large and very rapid adoption.

1.4. The 1970s: the infancy age of RFID

The 1970s was probably the infancy age of RFID development. Universities, research centers and companies were very actively working on many aspects of RFID that overlapped with integration technologies, RF solutions, communications and a large diversity of application scenarios. One of the most salient advances is the work developed at the Los Alamos Scientific Laboratory dedicated to "Short-range radio-telemetry for electronic identification using modulated backscatter" that was demonstrated in 1975 by A. Koelle, S. Depp and R. Freyman. In 1971, M. Cardullo demonstrated a passive radio transponder with 16-bit memory that was presented to the New York Port Authority and patented two years later (US Patent 3,713,148 in 1973). The patent of Cardullo was applicable for RF, sounds and light. The business model could be used in numerous applications such as car ID, toll, banking, credit cards, security, access control and medical surveillance. The Port Authorities of New York and New Jersey also tested systems for transportation developed by companies such as General Electric, Westinghouse, Philips and Glenayre. Although the tests provided positive results, the first commercially successful transportation application of RFID, electronic toll collection, was not yet ready [LAN 05]. During the same period, several companies developed RFID devices such as Raytheon's "Raytag" in 1973. RCA and Fairchild were active in several developments. R. Klensch from RCA developed an "Electronic identification system" in 1975, and F. Sterzer also from RCA developed an "Electronic license plate for motor vehicles" in 1977. Furthermore, T. Meyers and A. Leigh from Fairchild developed a "Passive encoding microwave transponder" in 1978 [LAN 05].

It appears that the 1970s were characterized primarily by the exploration of several implementations of RFID in real different scenarios and applications. Most of the investigated solutions were considered in applications such as animal tracking, vehicle identification as well as factory automation. Animal tagging was the major interest in Europe. However, due to the electromagnetic characteristics of biological tissues (very high permittivity and high losses), it was possible to implement only inductive systems. Examples of tags implanted into pets are shown in Figure 1.4. During the same decade, transport applications exploited the work on microwave systems developed at Los Alamos and by the International Bridge Turnpike and Tunnel Association (IBTTA) and the United States Federal Highway Administration. In the late 1970s, new applications began to appear and the number of interested companies, individuals and institutions working on RFID began to increase very rapidly. The potential of RFID technologies and the benefit they provided in terms of application monitoring as well as return on investment became obvious.

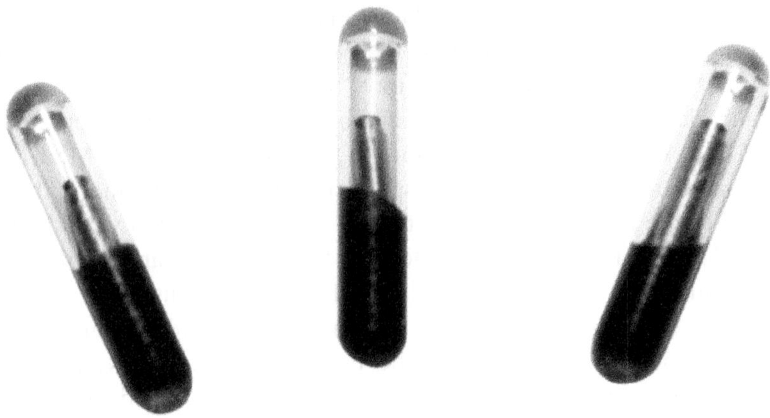

Figure 1.4. *Examples of implanted tags for pet tracking. These tags are based on inductive systems at LF (low-frequency) and HF (high-frequency) bands to take into account the electromagnetic characteristics of biological tissues*

1.5. The 1980s and 1990s: implementation of RFID

The very practical developments initiated in the 1970s continued to progress in the 1980s and the interest in RFID technology rapidly grew in various regions of the world. While in the United States, the major interest was in the transportation systems and access control; in Europe, the major applications included short-range systems for animals, industrial and business applications. Also, toll roads in some European countries were equipped with RFID solutions. The first commercial application for collecting tolls began in Norway in 1987 and was followed quickly by the Dallas North Turnpike in the United States in 1989. So RFID became a more profitable technology and a large number of new players arrived at this very rapidly evolving market.

The 1990s was a significant decade for RFID implementation, particularly in tolls and transportation systems in many countries around the world. Some of these implementations integrate real innovations. Among the significant innovations was the one introduced in the northeastern United States, where seven regional toll agencies formed the E-Z Pass Interagency Group (IAG) in 1990 to develop a regionally compatible system for toll collection based on RFID solutions. Such a system allows the use of a single tag and single billing account per vehicle to access highways of several toll authorities. Another remarkable innovation, the toll system of the highway which was opened in Oklahoma in 1991, allows vehicles to pass collection portals at highway speeds. A combined toll collection and traffic management system was installed in the Houston area in 1992. Later on, tolling and rail applications also began to appear in many countries including China, Hong Kong, Japan, Malaysia, the Philippines, Thailand, Singapore, South Korea, Australia, South Africa and various countries in Europe.

In the automobile sector, a remarkable application was the development of the Texas Instruments (TI) TIRIS system, which was used for the control of vehicle engines. The application of the TIRIS system was further extended to dispensing fuel, gaming chips, ski passes, vehicle access and so on.

During the 1990s, several European countries were actively developing RFID technologies at both hardware and application levels. In particular, semiconductor companies were active in the development of RFID circuits. Companies such as Philips (now NXP), Electronic Marin and ST-Microelectronics proposed RFID chips for LF, HF and UHF bands. During this decade, European countries paid more attention to the implementation of innovative RFID solutions. Special interest was focused on both RF and inductive technologies, in particular, for toll collection, access control and e-commerce. Europe became very active in the RFID market with developments including Alcatel, Bosch, STMicroelectronics and Philips. A pan-European standard was necessary for tolling applications in Europe, and many of these companies were working on the CEN standard for electronic tolling. Some examples of toll tags are shown in Figure 1.5.

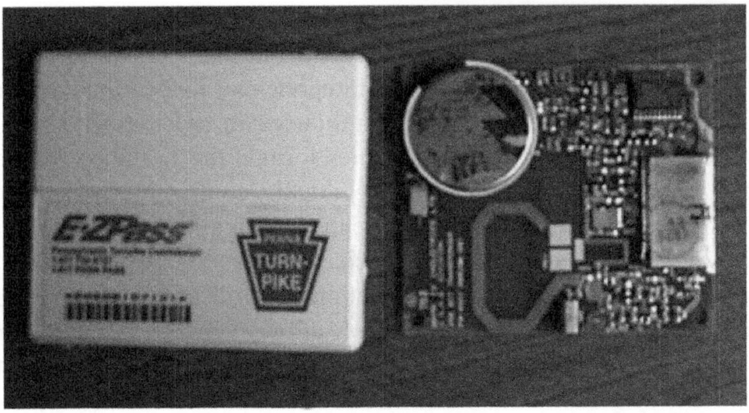

Figure 1.5. *Examples of tags for toll applications. Toll tags are based on RF technologies and in most cases are powered by a battery*

1.6. RFID chip age

Until the 1990s, RFID tags were composed of a PCB (printed circuit board) that contained electronic devices including a memory and antenna to ensure the wireless communication with the reader. Figure 1.6 shows some tags based on the PCB. Research and development in microelectronics and computer science did not slow down during the 1990s, and remarkable technological advances were accomplished that would benefit RFID. Indeed, for the first time, useful microwave Schottky diodes were fabricated on a regular CMOS integrated circuit. This advance in technology is a very important step because one of the major RFID circuit blocks is the rectifier that converts the RF in the interrogation signal into DC voltage to power up the RFID circuitry. Thus, it becomes possible to integrate all the functionalities of the RFID circuit in the same chip. This development allowed the construction of very compact RFID tags that contained only a single integrated circuit, a capability previously limited to inductively coupled RFID transponders. Large companies that were active in this evolution were IBM, Micron, Philips, STMicroelectronics and many other spin-offs.

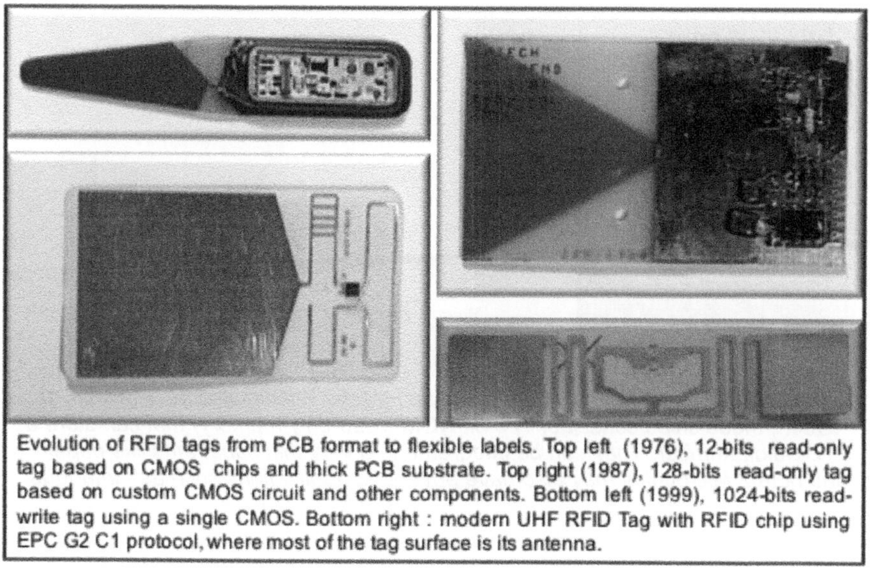

Figure 1.6. *Evolution of tags from the PCB circuit to the RFID chip. First tags used PCB and a suitable external package. Actual tags with reduced size have a label format with only two elements: RFID chip and antenna*

With the growing interest in RFID with respect to asset management and item traceability, and the opportunity for RFID to potentially positively substitute the barcode, it becomes difficult in the later part of this decade to count the number of companies who have entered the RFID market and developed many application scenarios.

1.7. Maturation of RFID

Integration technologies continued to rapidly advance at the beginning of the 21st Century, and some remarkable RFID chips were proposed by major companies. One of the most significant examples of technological process was the microchip developed by Hitachi, as shown in Figure 1.7.

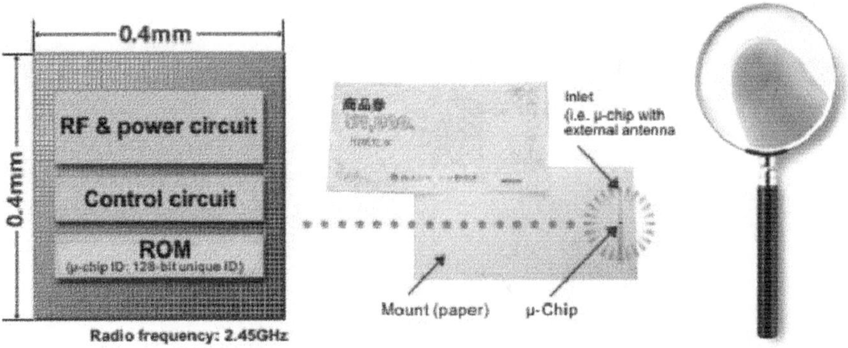

Figure 1.7. *The microchip developed by Hitachi. Known also as "Powder" or "Dust", these chips consist of 128-bit ROM (read-only memory) that can store a 32-digit number. They can be integrated into very thin substrates like paper and notes. Microchips may also have advanced applications such as "smartdust"*

This chip has a very limited size of 0.4×0.4 mm^2 and operates at the ISM (Industrial, Scientific and Medical) band of 2.45GHz. It exhibits 128-bit ROM. It is batteryless, decodes its microchip ID and transmits it back, and each μ-chip stores a unique 128-bit ID. The μ-chip has a read range of 25 cm with an external antenna. This evolution in terms of integration capabilities allows the building of tags with only two components: a single CMOS integrated circuit and an antenna. This development allows a reduction in the size of circuitry and the cost of tags, and an increase in functionality and reliability. Therefore, tags have become sticky labels that can be easily attached to objects to be managed somewhat similarly to the optical barcode. This similarity has pushed several research projects and partnerships between universities and companies to consider the adoption of RFID as a smart and better solution for traceability and logistics compared with the optical barcode. More recently, Hitachi has developed the world's smallest and thinnest RFID chip. Measuring only 0.15×0.15 millimeters in size and 7.5 micrometers thick, the wireless chip is a smaller version of the previous record. The new chip has a wide range of applications, from transportation and logistics to military and even consumer electronics.

The other salient example that remarkably illustrates the decisive advances in technology and new capabilities offered by RFID is the implanted tag introduced by VeriChip Corp, as shown in Figure 1.8.

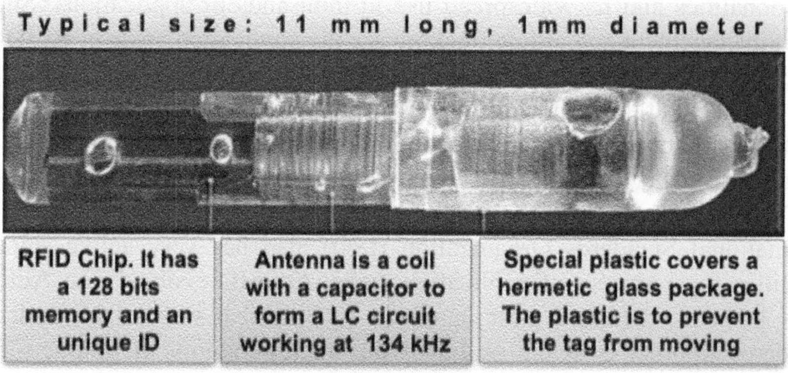

Figure 1.8. *Implanted tags developed by VeriChip Corp. The tag consists of an RFID microchip, a capacitor and a loop antenna wrapped around a ferrite core. It is enclosed in medical-compliant glass and coated in a substance called Biobond to avoid the migration of the tag within the body. The implant has the size of a grain of rice*

This is a 134.2 kHz RFID chip that can save lives and possibly limit injuries resulting from errors in medical treatments, as claimed by VeriChip [SID 17]. It received FDA (US Food and Drug Administration) approval in 2004. VeriChip suggests that the chip can be inserted in the rear part of the triceps of the right arm under the skin. According to VeriChip, the chip does not contain any medical records, but its 16-digit number could be linked to a database of patient medical information. When the tag is scanned, the number can be quickly cross-referenced to reveal specific medical data about the patient. The tags are similar to those being embedded in livestock and pets as an effective solution to tracking the mad cow health condition.

Apart from the previous remarkable examples of the technology outcome, technological development allows a reduction in the size of circuitry and the cost of tags. In addition to the RFID tag hardware, signal processing techniques and computer science allow the expansion of RFID functionalities and its widespread use in thousands of applications. Supply chain management and article tracking are the application areas of RFID that have grown rapidly. The dedicated facility Auto-ID center has been set up and organized at the Massachusetts Institute of Technology to bring together RFID manufacturers, researchers and users to develop standards, perform research and share information, in particular, for supply chain applications. EPC Global has assumed the task of standards for this application area. The International Standards Organization (ISO) also has very active standard activities for a variety of application areas. Auto-ID evolved into Auto-ID Labs in 2003, the same year in which EPC Global was established. The standard adoption is at the origin of the interoperability of products from different manufacturers, which results in increased compatibility, better performance and improved reliability of UHF RFID systems. Later, global standards, such as EPC Class 1 Gen2, encouraged major organizations, such as Walmart, Tesco, the US Department of Defense in the USA Decathlon, Inditex and Uniqlo, among other big retailers worldwide, to request their suppliers to become strictly RFID compliant with all their products. In the 2010s, major companies such as Macy's, Hudson's Bay Company, Marco Polo and a number of others turned the RFID technology into a commodity, especially in retail and industrial applications. Today, it is certain that there is no stopping RFID. It has already become an important part of our everyday life that enables many applications such as ticketing, e-papers, payments and access control.

The full potential of RFID is also being realized in other areas such as application software, development of privacy policies and other legal aspects. Besides, now that RFID has truly entered the mainstream, the development of the supporting infrastructure to design, install and maintain RFID systems is of high importance. As a matter of fact, the choice of a tag for a specific application first requires the definition of a set of parameters and criteria, and then it will introduce the infrastructure to enable the application. Figure 1.9 gives a possible classification of RFID tags using different parameters.

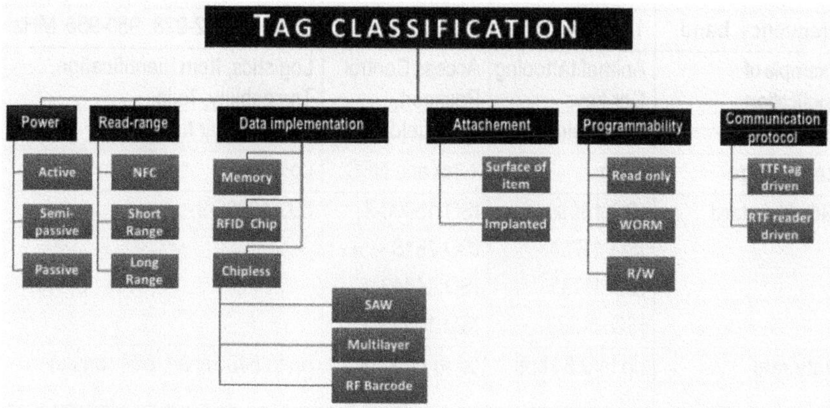

Figure 1.9. *Classification of RFID tags following different parameters. Several duties should be considered when selecting a tag category for an application. Due to their low cost, batteryless feature and long read range, passive UHF tags are very often selected*

1.8. Internet of Things: the next RFID frontier

RFID technology is a well-established technology that has been implemented in thousands of applications worldwide. One of the decisive advantages that RFID guarantees to its potential users is perhaps its standardized character which is of great importance. This standardization allows the interoperability between RFID devices from different manufacturers and geographical regions. Moreover, the passive feature of RFID tags is also very effective in terms of application reliability and the quasi-unlimited lifetime of tags. As pointed out in Figure 1.9, there are several categories of tags to fit the criteria and specific needs of certain applications. The most exploited categories of tags are reported in Table 1.1, which correspond to passive tags.

	LF 	HF 	UHF 
Frequency band	125-134 KHz	13.56 MHz	865-868, 902-928, 950-956 MHz
Example of application	Animal tattooing Car keys	Access Control Passport,	Logistics, Item Identification, Traceability, Tolls
Field zone	Near field	Near field	Near and far field
Read range	< 1 m	< 1m and NFC	up to 25 m
ISO Standard	ISO 18000-2 ISO 11784 ISO 11785 ISO 14223	ISO 18000-3 ISO 7618 ISO 14443 ISO 15693	ISO 18000-6
Data rate	up to 9.6 kbps	up to 64 kbps	up to 640 kbps (1500 tags/s)

Table 1.1. *The most exploited tag categories. These tags are passive devices that can be used for short range (less than 1 m) and long range (up to 28 m). Depending on the constraints of an application, the tags can be inserted into a suitable package before their attachment or integration to the item to be tracked*

While the frequency bands for LF and HF tags are the same worldwide, it is not the same for the UHF band. Indeed, there are three different bands: 866–868 MHz, mostly used in Europe and Africa; 902–928 MHz, mostly used in the Americas and some countries in Asia; and 950–956 MHz, mostly used in the Asia-Pacific region. Due to its long read range of several meters, several research programs and technology developments have been dedicated to RFID at the UHF band. Since the very first RFID chip with a sensitivity of -8 dBm introduced at the end of the 1990s, the pace of developments in chip sensitivity has continued to progress. Indeed, a gain of more than 10 dB has been achieved. Table 1.2 provides the progress of chip sensitivity during the last decades. It is remarkable to see that the actual read range of 28 m is quite comfortable for the majority of indoor applications.

Year	Chip sensitivity (dBm)	Read range (m)
1997	–8	5.07
1999	–10	6.38
2005	–12	8.03
2007	–13	9.01
2008	–15	11.34
2010	–18	16.02
2011	–20	20.16
2014	–22	25.38
2017	–23	28.50

Table 1.2. *Evolution of RFID chip sensitivity. Theoretical read range considering: operating frequency at 868 MHz, reader power at 2W ERP, reader antenna gain at 2 dBi, tag antenna as an ideal dipole and perfect matching between the RFID chip and the tag antenna*

Tags can now be used in a variety of applications and domains such as traceability, logistics and access control. They have become ubiquitous in industry and our daily life (ticketing, payment, passports, car keys, etc.). It is also being largely implemented in the area of health (smart hospital), personal assistance and anti-counterfeiting [DUR 18]. RFID is a standardized technology with inherent advantages of standardization, identification, wireless communication and low cost of tags, thereby providing decisive practical benefits that drive new developments in terms of concepts and applications. However, in all these applications, the major duty of the tag is to provide the ID and achieve the identification only. The perspective of RFID in terms of new paradigms for distributed ambient intelligence and the Internet of Things (IoT) is increasingly considered by many universities and industries. The future looks very promising for this technology. The full potential also requires advancements in the evolution of the RFID tag to provide more than just identification. Passive UHF tags using chips that require only –23 dBm for activation exhibit read ranges of more than 28 meters. They should allow the implementation of the last few meters of the IoT. To do so, the tags must be expanded in terms of functionalities, in particular sensing. The transformation of RFID tags into sensors is very effective in terms of wireless sensor networks. Indeed, the use of the standardized RFID communication protocol will allow the designers to exploit the normalized RFID commands to define new commands more specific to the sensing capability of the transformed tag. Moreover, like any other sensor, the calibration of the device is an important and critical issue. Such an objective can be reached by the association of some reference devices that provide correction signals to the interrogation system [MAR 10]. All these tasks will become possible if the cost of tags is low and the interrogation system is enriched and updated by additional commands to collect the sensed parameters.

There are several ways to transform an UHF RFID tag into a RFID sensor [TED 16]. They can be grouped into three categories. The first one will exploit the sensitivity of the tag antenna to the physical change that appears in its near-field region. The second category is more related to the behavior of the RFID chip and the variation of its electrical response as a function of some external parameters such as RF power and temperature. The third category consists of the integration of an external sensor to the RFID chip. It should be noted that the matching between the tag antenna and the RFID chip is the most relevant parameter that governs the performance of RFID tags, in particular the read range. Once the impedance states of the RFID chip are known, the designer should optimize the tag performance by properly choosing the antenna impedance that corresponds to a certain antenna topology on a well-defined support. Changing anything in the antenna topology, or in its immediate vicinity (near-field region), will turn into a measurable change in the performance of the tag-to-reader communication. This change can be observed in several communication parameters such as activation power of the tag, read range, frequency tuning, phase rotation and group delay. In the literature, several attempts and examples have been reported to design RFID sensors by exploiting this concept [TED 16]. The exploitation of the perturbation of the tag antenna by a metal plate to measure the inflexion of a bridge has been demonstrated. Sensing could also be achieved by adding a highly sensitive material to the antenna; that is, by adding carbon nanotubes (CNT), the gas concentration can be sensed. However, the exploitation of these RFID sensors in real applications has not been discussed much in the literature. Most of the published papers have focused on the concept of the device and restricted their study to validating the proposed structure by some experiments to demonstrate the effective variation of the backscatter signal with the sensed parameters. A real application was proposed in [SON 13]. It explained the transformation of an UHF RFID tag to detect the contamination of meat. The proposed device exploited the evolution of the permittivity of meat as a function of time. So, the measured RSSI was time dependent, corresponding to approximately 72 hours, which was considered as the optimal contamination time.

The RFID chip ensures several tasks; in particular, it scavenges wireless power from the reader and generates the DC voltage to empower the chip circuitry. Intensive work is dedicated to the optimization of chip characteristics in order to realize new functions of RFID tags. Energy harvesting for external sensors and channel diversity are among the most interesting capabilities. These two features are mainly governed by the non-linear behavior of RFID tags, which is primarily dependent on the characteristic of the scavenging section of the RFID chip. Spectral analysis and measurement of real backscatter signals demonstrated the existence of spectral components at the second and third harmonics of fundamental frequency. The exploitation of non-linear characteristics in the context of wireless communication was considered for localization applications in [HUG 09]. In RFID technology, the exploitation of such non-linear behavior opens the door to diverse new functions that can be easily integrated into the tag, especially by adapting the design of the tag antenna. This concept will be discussed in the following chapters.

1.9. Summary

In the 2010s, major companies turned RFID technology into a commodity, especially in retail and industrial applications. It has already become a part of our everyday life, and we use it without knowing that. In 2014, Smartrac, Impinj, Google and Intel joined forces to establish the RAIN alliance. The main objective of this alliance, announced in April 2014, is to boost the global adoption of UHF RFID technology, in order to promote awareness, increase education and support the universal adoption of UHF RFID technology. The members of the RAIN alliance are manufacturers, distributors, resellers and researchers who work with the EPC Gen2 UHF RFID specification, incorporated into the ISO/IEC 18000-6C standard. RAIN RFID is a wireless technology that connects billions of everyday items to the Internet, enabling businesses and consumers to identify, locate, authenticate and engage each item. On top of these advances, augmented tags are highly necessary. Exploiting the characteristics of the non-linear RFID chip is one of several ways to transform tags into augmented tags, providing the identification more surely.

At first glance, the concept of RFID and its application seems simple and straightforward. However, in reality, the opposite is true. Indeed, RFID is a solution that needs plenty of technologies, from systems engineering and software development to materials technology and mechanics. Other domains, such as microelectronics, circuit theory, antenna theory, radio propagation encryption and network engineering, are also necessary. The number of engineers involved in the development and application of RFID is increasing rapidly. This evolution will most likely continue to increase, allowing RFID to have a bright future.

2

RFID Technology: Main Principles and Non-linear Behavior of Tags

The objective of this chapter is to remind us of the principles of RFID technology with a focus on passive UHF RFID. It also details RFID studies and applications based on harmonic generation and non-linear behaviors.

2.1. RFID: a multilayer vision

RFID (radio-frequency identification) technology is a wireless data collection technology that is widely used in many applications and services such as logistics, manufacturing, access control and security, both in industrial and private environments [KLA 03, DOB 08, DOM 10]. Nowadays, the scope of passive RFID technology is not only limited to identification because tags are equipped with new capabilities that open the field for new applications dedicated to switching, monitoring and/or controlling purposes. Thus, the type of RFID applications is strongly evolving, for instance with tags capable of sensing environmental and human body parameters [MAR 10, NAZ 12, YVA 12], or tags able to communicate and share information with other tags [NIK 12]. A meaningful indicator of this trend is the amount of research efforts made to study and present RFID technology as one of the frontrunners in the new paradigm of the Internet of Things (IoT) [LAN 08].

RFID technology and its evolution can be better explained by placing each of its technical components in a modular and multilayer stack model, as shown in Figure 2.1. The multilayer model uses the Open Systems

Interconnection (OSI) terms and is compatible with the ISO 18000-1 [ISO 15]. It provides parameter definitions for globally accepted frequencies in RFID [ISO 12] and is also valid for positioning emerging RFID technologies such as chipless RFID in where the basic identifier is obtained from the electromagnetic signature of a resonator without a silicon chip [PRE 09, TED 13]. The model is composed of three layers: the application layer, the communication layer and the physical layer [GIL 05].

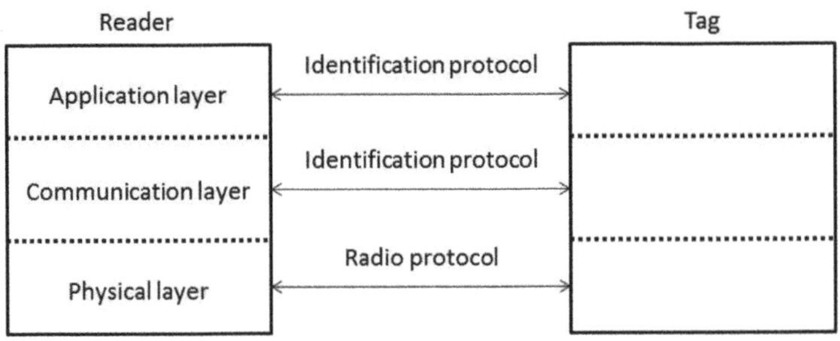

Figure 2.1. *Air interface protocol stack*

The application layer represents information to be handled by the user. This could be an identifier set as the filter in the reader, which will allow the detection of a specific tag whose information can be read or written into a database (e.g. the title of a book). In this layer, protocols and rules for end-user software tools are positioned. In the case of the chipless RFID system, the rules that format the information obtained from the translation of signal processing into exploitable end-user data could be defined here.

The communication layer represents the protocols that rule the communication between the reader and the tag. Collision avoidance protocols are defined in this layer. When drawing an analogy with a chipless RFID system, given the absence of smart components on the chipless tag, the basic definitions of collision avoidance could be split into its physical and application layers.

The physical layer represents the air interface (i.e. frequency, operating channels, modulation, data encoding, timings and bit rate), circuit connections, powering, signal strengths, chip architecture (for readers and tags), structural connections (between the antenna and the chip), structural shapes (responsible for the radio frequency signature in chipless RFID), resonances and polarization. The features of this layer define whether the tags work in the LF (low-frequency), HF (high-frequency), UHF (ultra high-frequency) or microwave band, and also determine its linked protocols in upper layers. Regardless of the frequency band, most RFID systems only use a few kHz to MHz of bandwidth and are thus classical narrow band systems.

The EPC Global Class-1 Gen-2 UHF RFID standard (ISO 18000-6) defines the operations that rule the communication between reader and tag. Indeed, its scope covers the above explained three layers of the protocol stack.

For instance in the physical layer, the standard defines the radar cross section and read range based on link frequencies, antennas gain, polarization, free space path loss, etc.

In the communication layer, the standard defines the reader-talk-first sequence of messages, the coding mechanisms (which determine the data rate or bit duration) and also the anticolision mechanism based on the ALOHA algorithm.

The evolution of the standard from version 1 (Gen2V1) to version 2 (Gen2V2) in 2014 have pushed manufacturers to add new functionalities to the chip such as crypto-security or file management. These new functionlities will trend to fill the application layer of the proposed protocol stack model.

Following the multilayer RFID model, this chapter is concerned with the physical and communication layers of the EPC Global Class-1 Gen-2 UHF RFID standard, with a focus on RFID chips. It mainly deals with the effects intrinsically produced in consequence of the non-linear behavior of RFID chips. In order to unfold the features presented in the following chapters, the next two sections will break down the theory and definitions of passive UHF RFID and non-linear RF networks.

2.2. Focus on passive UHF RFID technology

2.2.1. *Working principle*

In passive UHF RFID, tags that are compliant with the EPC Global Class-1 Gen-2 standard (adopted as ISO 18000-6C), do not have an autonomous energy source like a battery, but obtain it via electromagnetic fields transmitted by the reader. The general working principle of passive UHF RFID systems is shown in Figure 2.2. The reader generates a carrier wave that is transmitted by the reader antenna. Reader antennas are typically directive, i.e. they only illuminate a certain volume, called the read zone. If a tag is inside this read zone, the power transmitted by the reader activates this tag and makes it ready to receive commands. The EPC Global Class-1 Gen-2 is a reader talks-first protocol, i.e. the tags wait until they are addressed. On receiving a command transmitted by modulating the carrier wave (CW), the tag sends its identification code or parts of its memory. The tag does not actively transmit this data, but reflects part of the incident reader carrier wave by deliberately de-tuning its own antenna that generates two different radar cross sections (RCS), as shown in Figure 2.2. This principle of modulating the reflected power is known as backscatter modulation [NIK 06].

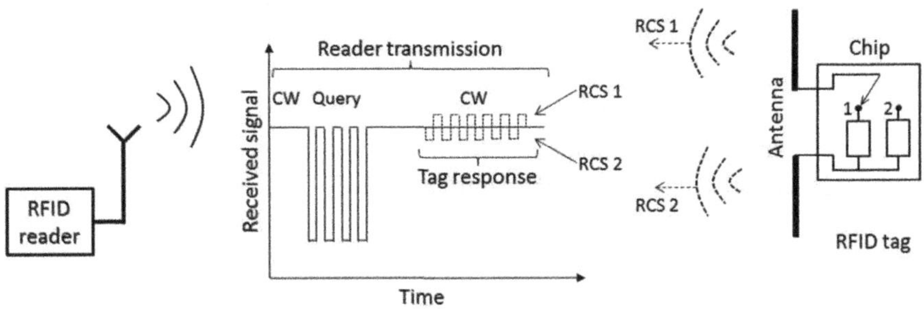

Figure 2.2. *Working principle of passive UHF RFID systems*

2.2.2. *Reader*

Based on its design and the technology used, the RFID reader (also called the interrogator) may be a reader-only or a reader- and writer-device. A practical example is shown in Figure 2.3. A reader typically contains a radio frequency module (transmitter and receiver), a control unit and a coupling element with the tag (i.e. the reader antenna). In addition, many readers are fitted with an additional interface (e.g. Ethernet, RS-232) to enable them to forward the data received by another system (e.g. PC, robot control system).

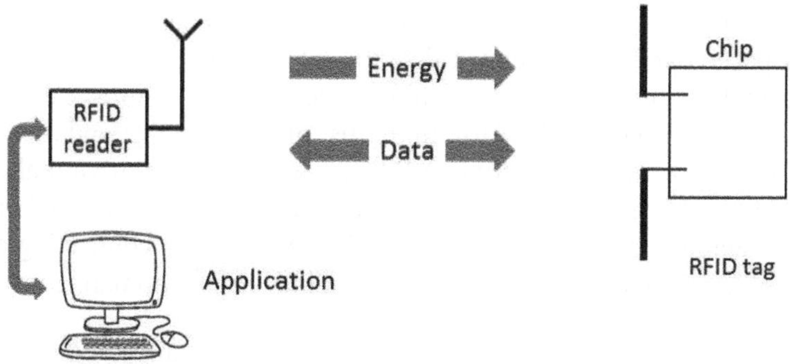

Figure 2.3. *Reader and tag in a passive UHF RFID system*

2.2.3. *Tag*

With regard to the aimed application, there is a multitude of tags with different shapes and sizes. However, a common denominator is that all of them are composed of three main components: (i) the substrate or cover support (i.e. the antenna cover or support); (ii) the antenna itself whose choice depends largely on the application environment (in terms of both ergonomics and performance); (iii) a dedicated chip (integrating basic processing functions, memory, and lately sensor and security capabilities [WIR 17, UCO]). They are collectively called the RFID tag, label or inlay.

2.2.3.1. *Tag antenna*

Tag antennas used for most applications in UHF RFID systems consist of a printed or etched metal surface on top of a thin and flexible sheet of plastic or paper. Tags are often equipped with an adhesive film that can be attached to flat surfaces. Many shapes of such antennas can be found, but technically most of them are dipole antennas which have meandered conductors and a widened metallization at dipole ends that function as an end capacitance [MAR 08]. This technique allows the sizing down of the antenna compared to the half-wavelength dipole antenna. Tags based on dipole structures are linearly polarized with an omnidirectional radiation pattern in one plane. To match the impedance of shortened dipole antennas with the input impedance of the chip, several techniques have been proposed. The most common technique is a inductively coupled loop or slit implemented in the center of the dipole [DAT 12]. The length of such antennas ranges between 5 cm and 15 cm. For even smaller tags, a metallization in the form of a loop is used.

An innovation in tag side was introduced in 2017. It consists of embedding a slender passive RFID tag in a textile yarn, which is called the RFID yarn. A gamma-match impedance matching technique ensures the sound performance of the RFID yarn in terms of read range and bandwidth, maintaining a slender geometry [AND 17]. This new form of yarn allows the RFID tag to be embedded in different objects and easily conformed to different shapes. This shows passive RFID's potential in connecting objects, for example in smart textiles, wearables and industrial automatization.

From an application point of view, tag antennas can be considered the strategic point in the radio transmission chain of an RFID system. Except for the RFID yarn, it is preferable if they are shortened, which inevitably comes at the cost of low bandwidth and poor radiation efficiency. Furthermore, they are attached to objects that may contain fluids, metal, dielectric materials or materials with high dielectric losses. Such objects often require specialized antenna designs or may even need a radome to ensure a distance around the tag. The reason for this is that electrically small antennas are more prone to proximity effects. Materials in the proximity of the tag antenna cause a shift in center frequency and input impedance, often referred to as de-tuning [DOB 15]. This effect impairs the matching between the tag antenna and the chip, which, in turn, reduces the power available to operate the digital circuits of the chip. Moreover, lossy materials in the proximity of the antenna will absorb power. Figure 2.4 shows a parametric study of the RFID tag reflection coefficient when varying its distance to a metal plate. There are drastic changes in the electrical behavior of the antenna when the tag gets closer to the metal plate. Therefore, proximity effects due to tagged objects on tag performance should be considered in the antenna design process for each application case.

However, with the objective of providing a multi-standard operation, for instance HF RFID combined with UHF RFID, UHF-Europe (866–868 MHz) with UHF-USA (902–928 MHz), UHF RFID with microwave RFID (2.45 GHz and 5.2 GHz) or UHF RFID with NFC, the RFID tag antenna (and also the reader antenna) can be dual band antennas. A multiband operation can be achieved in a traditional way by using several resonant elements, such as PIFA structures [HIR 06], slot antennas [JEO 06, MIN 10], multilayer configuration [MOB 10], spiral coils [LON 12, MAY 08, PAR 10] or asymmetric folding dipole [MAL 11]. It should be noted that all dual band tag antennas for HF–UHF bands or UHF–microwave are proposed in a claimed scenery of a multi-standard RFID operation that does not exist yet. In this scenery, one chip is expected to be able to support more than one protocol of communication. Some patents have already protected this idea [MAY 12] and, for instance, the commercial chip EM4423 of EM Microelectronic contactless devices brings innovative features to the NFC and EPCTM worlds.

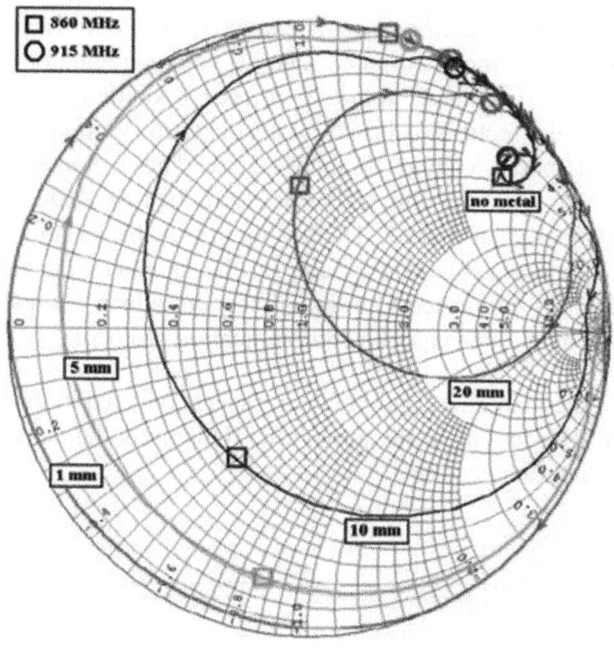

Figure 2.4. *Simulated RFID tag reflection coefficient when varying its distance to an infinitively extended metal plate [DOB 05]. For a color version of this figure, see www.iste.co.uk/andia/non-linearites.zip*

2.2.3.2. *Passive chip*

Passive RFID tags do not have an internal battery, and therefore they must draw the required power for their operation from the electromagnetic field transmitted by the reader. The RF energy radiated by the reader is used to both supply the digital section of the tag and allow data transmission from the tag to the reader by modulating the backscattered radiation [DEV 05].

The architecture of a passive UHF RFID tag is shown in Figure 2.5. The coupling element is an antenna, which is typically a dipole. An impedance matching network between the antenna and the RFID chip ensures the maximum power transfer between the two elements in one of the impedance states of the chip. A rectifier circuit converts the input alternating voltage into a DC voltage, which is used by a series voltage regulator to provide the required regulated voltage for the correct operation of the tag. The architecture of the rectifier is based on a Cockcroft–Walton circuit with two or more diode-based voltage doubler stages. These diodes determine the non-linear behavior of the chip [DEV 05, BOL 10]. An ASK demodulator is primarily composed of a peak detector circuit. A backscatter modulator is used to modulate the impedance seen by the antenna during transmission. The RF section is connected to a digital section, which is typically a finite-state machine able to manage the communication protocol.

Note that the theoretical and practical aspects covered in this book mainly consider the RF sections of the passive RFID tag, i.e. the antenna, matching network and rectifier sections.

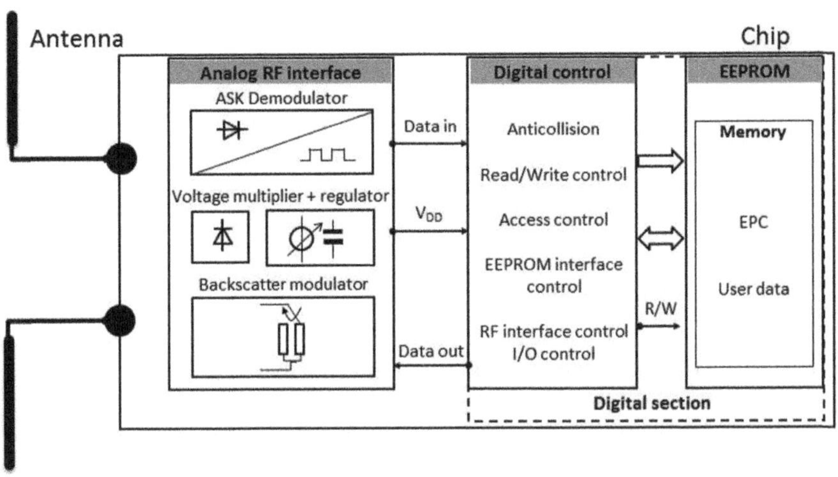

Figure 2.5. *Architecture of a UHF RFID passive chip*

2.3. Non-linear RF networks and harmonic generation

2.3.1. *Effects of a non-linear device*

All realistic RF devices have at least small transmission losses. An ideal linear component does not exist in practice because all realistic devices are non-linear at very low signal levels due to noise effects. In addition, practical components may also become non-linear at high signal levels. In the case of active devices, such as diodes and transistors, this may be attributed to effects such as gain compression or the generation of spurious frequency components due to device non-linearities, but all devices ultimately fail at very high power levels. In either case, these effects set a minimum and maximum realistic power range, or a dynamic range, at which a given component or network will operate as desired [DAV 05]. It is this non-linearity that is of great utility for desirable functions such as amplification, detection and frequency conversion [HIN 84]. However, non-linear device characteristics can also lead to undesirable effects such as gain compression and the generation of spurious frequency components. These effects may further lead to increased losses, signal distortion and possible interference with other radio channels or services. Some of the many possible effects of non-linearity in RF and microwave are as follows [MAA 03]: harmonic generation (i.e. multiples of the fundamental signal), saturation (i.e. gain reduction in an amplifier), intermodulation distortion (i.e. products of a two-tone input signal), cross-modulation (i.e. modulation transfer from one signal to another), amplitude and phase conversion (i.e. amplitude variation causing a phase shift), spectral regrowth (i.e. intermodulation with many closely spaced signals).

2.3.2. *Theory on the effects of a non-linear device*

Figure 2.6 shows a general non-linear network with an input voltage v_i and an output voltage v_o.

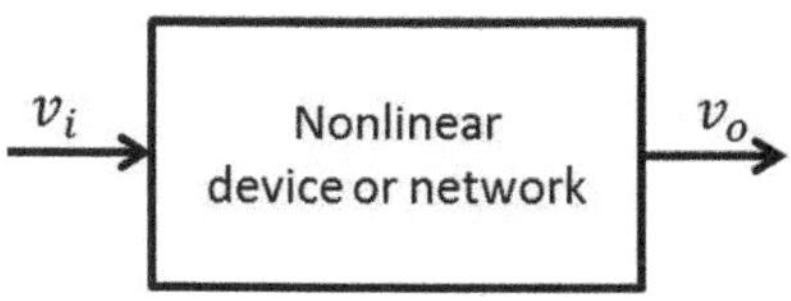

Figure 2.6. *Non-linear device or network*

In general, the output response of a non-linear circuit can be modeled as a Taylor series in terms of the input signal voltage:

$$v_o = a_0 + a_1 v_i + a_2 v_i^2 + a_3 v_i^3 + \cdots \quad [2.1]$$

where the first three Taylor coefficients are defined as:

$$a_0 = v_o(0) \qquad \langle \text{DC output} \rangle$$

$$a_1 = \left.\frac{dv_o}{dv_i}\right|_{v_i=0} \qquad \langle \text{linear output} \rangle$$

$$a_2 = \left.\frac{d^2 v_o}{dv_i^2}\right|_{v_i=0} \qquad \langle \text{squared output} \rangle$$

Different functions can be obtained from the non-linear network depending on the dominance of particular terms in the expansion. The constant term, with coefficient a_0, leads to rectification, converting an AC input signal to DC, which is the base of the powering technique in passive RFID tags. The linear term, with coefficient a_1, models a linear attenuator when $a_1 < 1$, or an amplifier when $a_1 > 1$. The second-order term, with coefficient a_2, can be used for mixing and other frequency conversion functions.

Practical non-linear devices usually have a series expansion containing many non-zero terms, and there will be a combination of several of the above effects. In accordance with the topic of this book, the next section will pay special attention to harmonic generation phenomena.

The traditional way of demonstrating how new frequencies are generated in a generic non-linear network (Figure 2.6) is to describe the behavior of the network by considering the case where a single-frequency sinusoid is applied to its input:

$$v_i = A cos \omega_0 t \quad [2.2]$$

In this case, the Taylor series give the output voltage as:

$$v_o = a_o + a_1 A cos\omega_0 t + a_2 (A cos\omega_0 t)^2 + a_3 (A cos\omega_0 t)^3 + \cdots \quad [2.3]$$

$$v_o = \underbrace{\left(a_o + \frac{1}{2} a_2 A^2\right)}_{dc} + \underbrace{\left(a_1 A + \frac{3}{4} a_3 A^3\right) cos\omega_0 t}_{Fundamental}$$

$$+ \underbrace{\left(\frac{1}{2} a_2 A^2\right) cos2\omega_0 t}_{2nd\ harmonic} + \underbrace{\left(\frac{1}{4} a_3 A^3\right) cos3\omega_0 t}_{3rd\ harmonic} + \cdots \quad [2.4]$$

Consequently, it can be observed that a portion of the input signal at frequency ω_0 is converted to other frequency components. For example, the first term represents a DC voltage, which can act as a useful response in a rectifier application. Voltage components at frequency $2\omega_0$ or $3\omega_0$ can be useful for frequency multiplier circuits. In specific devices (e.g. amplifiers), the presence of other frequency components will lead to signal distortion if those components are in the passband. For a single input frequency (or tone), ω_0, the output will generally consist of harmonics of the input frequency of the form $n\omega_0$, for $n = 0, 1, 2, \ldots$. If these harmonics lie outside the passband of the device, then there will be no interference with the desired signal at frequency ω_0.

However, the situation is different when the input signal consists of two closely spaced frequencies [MAA 03]. The combination of two input frequencies applied to a non-linear network produces intermodulation products, which is the origin of intermodulation distortion. Let us assume a case in which a double-frequency composite signal is applied to its input:

$$v_i = A(cos\omega_1 t + cos\omega_2 t) \quad [2.5]$$

The output is composed of a DC term, a fundamental term and intermodulation products, which generate frequencies other than the harmonics of the two input frequencies:

$$
\begin{aligned}
v_o = {} & \underbrace{(a_o + a_2A^2)}_{dc} + \underbrace{\left(a_1A + \frac{9}{4}a_3A^3\right)(cos\omega_1 t + cos\omega_2 t)}_{Fundamental} \\
& + (a_2A^2)(cos(\omega_1 - \omega_2)t + cos(\omega_1 + \omega_2)t) \\
& + \left(\frac{1}{4}a_2A^3\right)(cos2\omega_1 t + cos2\omega_2 t) \\
& + \left(\frac{1}{4}a_3A^3\right)(cos3\omega_1 t + cos3\omega_2 t) \\
& + \left(\frac{3}{4}a_3A^3\right)(cos(2\omega_1 \pm \omega_2)t + cos(2\omega_2 \pm \omega_1)t) + \cdots
\end{aligned} \quad [2.6]
$$

Regarding the composition of different frequencies of the output signal of a non-linear network, several studies in antennas and circuits can be found in the literature [JAN 88, CAB 04, FON 82]. Some related works in the RFID field as well as in batteryless non-linear devices will be detailed in the next section.

2.4. Non-linear behavior and associated applications in the RFID field

2.4.1. *Measurement of backscattered harmonics*

The first experimental characterization of harmonic scattering from passive UHF RFID tags was published in 2009 by P. V. Nikitin *et al.* [NIK 09]. With some limitations on equipment (e.g. activating and measuring the RFID tag at the same time), the authors analyzed the first radiated harmonics of three commercial tags: AD-222, AD-223 and AD-224 [AVE 09]. The objective of the study was to quantify the signal levels radiated at harmonic frequencies in order to determine whether they had an impact on other RF systems deployed in the same environment.

The measurement setup used was the compact broadband GTEM (gigahertz transverse electromagnetic) cell TESCOM TC5060. Such cells are a type of electromagnetic compatibility (EMC) test chamber used for radiated EMC testing (e.g. radiation or susceptibility testing). They allow the creation of a controlled uniform field at a desired location, without considering the frequency-dependent responses of the transmitting antenna and the anechoic chamber material. The test consisted of injecting an RF signal at 880 MHz with 10 dBm power in order to generate an electric field inside the cell. An RFID tag was placed inside the cell and the backscattered field from the tag was detected with the sensing probe antenna connected to a second port. The probe antenna was a simple half-wavelength dipole printed on an FR4 substrate and located 10 cm away from the tag.

The results of the reported measurement are summarized in Table 2.1. The experimental results show the presence of harmonic signals with a relatively low power compared to the fundamental frequency. The power level for the second harmonic in the cell is constant with the presence or absence of the tag. On the contrary, the power of the third harmonic strongly increases as soon as a tag is present, which shows that tags can radiate third harmonic signals.

Harmonic power [in dBm]	No tag	AD-222	AD-223	AD-224
Fundamental frequency	–4	–6	–6	–6
Second harmonic	–72	–72	–72	–72
Third harmonic	–86	–63	–64	–68

Table 2.1. *Measured power of harmonics at a 10 dBm input signal*

This first type of experiment in the RFID field shows the remarkable ability of passive UHF RFID tags to generate and backscatter harmonics. At this level, it was possible to stipulate that a tag could backscatter harmonics and intermodulation products when interrogated with a modulated or unmodulated reader signal. However, it is worth noting that the proposed measurement setup did not consider an activated tag, i.e. no real tag response was measured except the harmonic multiples of the carrier wave sent by the RF transmitter and multiplied by the tag circuitry. Although this study aroused some interest in the RFID field, no further research efforts were made. This could be because the observed power levels were considered sufficiently low with no impact on traditional applications.

Chapter 3 will describe in more detail the complete experimental characterization and associated configuration setups of harmonic generation and radiation. In regard to this first approach, tags will be activated during the characterization steps, and complementary radiated and conducted measurements will be presented.

2.4.2. *Wireless sensor tags*

Regarding end-user sensing applications, two non-linear effects have so far been exploited in RFID or batteryless devices based on rectifying diodes. The first one is harmonic generation in tags, called frequency doubler tags, with its applications in detection. The second one is intermodulation distortion, with its applications in localization.

2.4.2.1. *Frequency doubler tags*

Frequency doubler tags, also known as one-bit RFID tags, are used to check and monitor the possible presence of an object (attached to a tag) in the interrogation zone of a reader. In this technique, the antenna (e.g. crossed dipole [ORE 11] or fractal loop [PSY 08]) is disturbed by the impinging radar wave and is connected to a small and light low-barrier Schottky diode. The current induced in the antenna is propagated through the diode that produces harmonic signals due to its non-linearity. The predefined harmonic signal (the second harmonic) is backscattered by the antenna, which is properly tuned for the two frequencies. This type of harmonic generation is reliably adopted in the microwave frequency range, leading to the harmonic radar concept [PSY 08, COL 04, ROS 12]. Such a signal can be detected by means of a microwave receiver without being masked by reflections from the surrounding scenario. The detection is considered feasible even if modulation is applied to the carrier. The envisaged applications are the harmonic radar for insect tracking [PSY 08], traffic applications [HAR 73] and avalanche rescues [ORE 11].

Additional works were recently presented based on the same concept, but whilst considering a near-field operation at 7.5 MHz and using a multiplier diode made of an organic material [MAR 14]. Pentacene diodes are organic devices, with pentacene used as a semiconductor. These diodes are used in harmonic tags to provide the non-linearity required to generate the harmonic. A parameter called conversion loss, defined as the difference (in dB) between the input power of the fundamental frequency and the detected harmonic power, is used in order to quantify the quality of the process. The best conversion loss for the harmonic tag of one bit is 52.5 dB, detected at a distance of 12.5 cm with an input power of 10 dBm. Since only one diode is used, the most significant harmonic generated is in this case the second harmonic. The major drawback of this auto-proclaimed eco-friendly chipless tag is the large size of the antenna (14 cm × 14 cm).

Another application that exploits the harmonic generation of non-linear devices is the detection of counterfeit RFID cards [ROM 09]. The principle is based on the electromagnetic characteristics of these cards rather than the digital information that they transmit. One of the steps considers the electromagnetic signature that RFID cards produce by radiating the harmonic frequencies of the communication carrier between the reader and the card at 13.56 MHz in the near field. The reported results show that different RFID card manufacturers can be automatically distinguished with the broadband analysis.

For wireless sensor applications in the cited references, it is worth noting that no work has been reported concerning the characterization of non-linear devices (i.e. diode-based rectifiers), either in impedance or in harmonic production. Details about the experimental characterization and modeling of generated harmonics from non-linear devices (and notably in RFID chips) will be discussed in Chapters 3 and 4, respectively.

2.4.2.2. *Intermodulation distortion*

So far, no specific application exploiting intermodulation distortion in totally batteryless devices has been found in the literature.

In 2009, Gomes *et al.* proposed the use of a semi-passive tag based on intermodulation distortion for localization scenarios [HUG 09]. The operating principle is shown in Figure 2.7. The reader sends an RF signal at the frequency f_1, modulated by a pseudo-random sequence and an unmodulated carrier RF signal at a close but different frequency f_1. When the signal reaches the tag, it is demodulated and re-modulated on a different carrier (built by the intermodulation product provided by the non-linear component, i.e. a diode) and re-emitted to the air interface at the frequency $(2f_1 - f_2)$. The modulation allows the identification of a specific tag in the analyzed environment. The reader has an RF receiver front-end tuned to the frequency $(2f_1 - f_2)$, which allows it to capture a replica of the transmitted signal. The two pseudo-random signals (transmitted and received) can be compared in time, thereby allowing the calculation of the travel time. Time delay indicates the distance between a reader and the considered tag. For finding the tag localization, at least three readers have to be used in order to apply the trilateration principle.

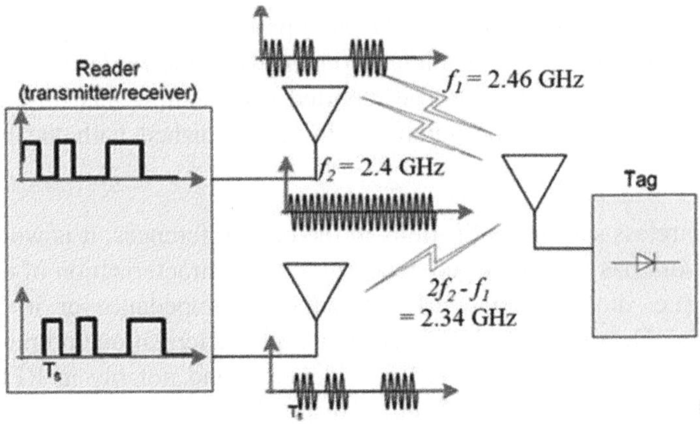

Figure 2.7. *Localization system exploiting the intermodulation phenomena*

2.4.2.3. *Multi-carrier system*

In order to increase the read range and overcome the problem of insufficient power in passive RFID tags, several authors have proposed a multi-carrier UHF passive RFID system composed of a reader transceiver, a carrier wave emitter (CWE) and a passive UHF RFID tag [MIN 11, HSI 07a, HSI 07b, HSI 11]. The proposed system offers additional energy to Class-1 Gen-2 tags and also generates frequency diversity on its modulated response by means of a multiple-carrier operation. The reader can be benefited from the frequency diversity gain in order to overcome the multi-path fading problem.

The proposed approach profits from the narrow frequency range operation of passive RFID tags using different carrier frequencies that are closer to each other. The operation is as follows: (i) the CWE located close to the tag, as a radio source, emits a carrier wave to illuminate the tags and provide them additional wireless power; the CWE frequency is approximately 10 MHz distance from the standard carrier wave of the reader transceiver; (ii) in addition, a transceiver carrying the logic of an RFID reader transmits and interprets the communication protocol with the tag at the standard frequency; (iii) special processing of signals sent from the reader to the tag is performed in order to mitigate the effect of additional carriers, which allows the tag to understand the commands.

The reported results show the tag response with a modulated backscattering signal for each carrier wave source (CWE and transceiver). This phenomenon is defined by the authors as frequency diversity, and further work is envisaged to combine the modulated backscattered response around each carrier. Although the reported results showed an improvement in the read range that is four times longer than ordinary Class-1 Gen-2 tags, there are some inconveniences introduced in the tag side, which can trigger demodulation errors of reader-to-tag commands. In addition, the processing changes the standard communication protocol for UHF passive RFID tags, thus making them unviable for traditional/commercial applications.

2.5. Summary

Nowadays, RFID has become a well-known technology standardized with commonly known and emblematic applications. Among the existing standards, UHF RFID is still evolving with numerous improvements for different layers (in terms of anti-collision protocols, memory size, chip sensitivity, antennas dedicated to harsh environments, etc.), as well as with new functionalities included in the tags in addition to the identification function. These additional functionalities have triggered the upgrading of the standard EPC Class-1 Gen-2 from version one (V1) to version two (V2) with a major focus on security and data protection. In this context, the chip UCODE DNA from NXP is the first one to apply crypto approaches to protect chip information. Furthermore, research and innovation in passive RFID does not stop and new functions proposed by chip manufacturers continue to push the standards for including and regulating this fast-growing technology. Hence, passive UHF RFID technology is still very promising, especially in the emerging area of IoT.

Non-linear effects such as harmonic generation and intermodulation in the tag side have been exploited as a one-bit system where the presence of harmonics or intermodulation components serve to detect the presence of tags. However, as in most RF systems, the non-linear properties of EPC Class-1 Gen-2 chips imply uncontrolled backscattered harmonic signals (especially the third harmonic), which *a priori* have undesirable effects such as energy loss. This loss is shown by signals transmitted at other frequencies than the operating frequency. Chapters 3 and 4 aim to complete and better understand this phenomenon with measurement and modeling studies, respectively.

3

Characterization Platforms for Passive RFID Chips and Tags

This chapter presents two specialized test platforms to characterize the performance of EPC Class-1 Gen-2 RFID chips (commercially known as RAIN [RAI 17]) with and without the tag antenna. The importance of these platforms is the measurement acquisition function that operates while the standard communication reader tag is established. With the aim of evaluating the non-linearity of those chips, specific analysis methods and equipment configuration are developed. This chapter unfolds the techniques and methods used for this evaluation.

These test platforms allow measuring, among other variables, the threshold activation power, chip impedance at both modulation states and parameters of the modulated chip response in both time and frequency domains. In addition, measurements can be set during sweeps of frequency, power and time. The added value using these platforms is its large bandwidth operation that allows performing measurement at frequencies other than the reader carrier, for instance the harmonic multiples.

3.1. Introduction

The operation principle of the passive RFID system was presented in Chapter 2. Concerning functionality, it can be summarized by two operations:

– Wireless powering of the tag, where the reader performs a wireless power transmission to the tag. The tag must collect this RF power and convert it into DC power to later consume in order to complete the communication operations linked to the data such as responding its ID by modulating its backscatter signal;

– Wireless data collection, described by a constant interaction between reader and tag, where the reader acts as a base station in order to read or write information from or to the tag, respectively.

The RF-to-DC power conversion task performed by the RFID chip is the well-known process of radio-frequency rectification, in which the chip harvests enough energy to be able to interact with the reader. The rectifier operation is possible thanks to the use of two or more diode-based voltage doubler stages (details of the rectifier architecture and operation will be discussed in Chapter 4). These diodes determine the non-linear behavior of the RFID chip [DEV 05, BOL 10].

The objective of this chapter is to observe the characteristics of the modulated backscatter signal and evaluate the information eventually carried by the harmonic signals, i.e. the consequence of the non-linear behavior of the RFID chip. The study is done when the communication from the reader to the tag is fully established. In addition, analysis of the reader harmonic production on the forward link is performed. An important contribution to ease the power spectral density (PSD) analysis of the backscattered signals, including harmonics is also presented. The contribution uses a pre-configured RFID air interface in order to produce a periodic sequence of bits on the tag response. A dependency analysis of the backscattered harmonic generation on different parameters such as reader frequency, reader power and periodicity of bit sequence is explained.

Section 3.2 explains the setups and measurement procedure of the tag response with special attention to the RFID air interface treatment. Section 3.3 presents the platform to characterize complete RFID tags, including the tag antenna, the so-called radiated characterization. Section 3.4 focuses on the characterization of RFID chips itself, without the tag antenna. This method, so-called conducted measurements, makes it possible to study the effect of the impedance seen by the chip on the harmonic generation. Finally, section 3.5 summarizes the milestones of the characterization platforms.

3.2. Measuring the backscattered tag response

3.2.1. *Harmonic backscattering*

The reason why the harmonic signals are re-radiated by the tag antenna can be explained by looking into similar technologies at low power rectification. The rectennas, devices formed by one antenna loaded by one rectifier circuit commonly used for electromagnetic energy harvesting (EEH) [MCS 92], have a similar rectifier architecture to passive RFID tags. Although rectennas used for EEH and passive RFID tags have a similar RF section, there is a small but crucial difference in the non-linear treatment. Because of the non-linearity behavior of the diodes used in the rectifier, appropriate techniques of simulation and optimization are used in those circuit designs. The harmonic balance (HB) is a simulation method that considers the effect of the harmonic signals produced by non-linear devices [GEO 10]. In the rectenna case, the harmonic currents flow to the DC section and to antenna terminals. If they are well filtered in the DC section side, a pure DC signal can be achieved at the output, and then an enhanced RF-to-DC conversion efficiency is achieved [JIY 04, ERE 12, GEO 10]. The phenomena can be simulated with the HB method, as it is possible to predict and to control the behavior of the complete rectenna, both in the DC section and in the antenna side. In passive UHF RFID chips, the harmonics reflected through the DC section are treated during the chip design process, and an optimal DC signal then feeds the digital section of the chip. However, the harmonics are reflected through the antenna since no harmonic treatment is performed on the RFID antenna design. The antenna design in passive RFID tags begins directly from the knowledge of one impedance value. The known impedance is that of the chips at the fundamental frequency provided in data sheets (e.g. [IMP]). Current design processes aim to ensure matching at the frequency of operation, but the antenna–chip impedance behavior around the harmonic frequencies is unknown. As long as the antenna and the chip are independently designed [DOM 10], and depending on the antenna structure, it can radiate some of the harmonic signals generated by the rectifier, which finally triggers in harmonic backscattering [NIK 09, AND 13].

3.2.2. *Measurement techniques*

The measurement procedure of backscattered harmonics is based on the observation of RFID signals in the time domain, and the calculation of its

PSD is performed by using fast Fourier transform (FFT). Figure 3.1 shows the conceptual procedure. The FFT calculation is made over a carrier wave (CW) portion after the Query command, as shown in Figure 3.2. The measurement procedure directly exploits the possibilities that the RFID air interface offers on its physical layer to facilitate the spectral analysis. The challenge is done by parameterizing the tag response to produce a cyclic and repeatable signal that facilitates the PSD analysis: repeatable in order to easily analyze and compare the results and cyclic in order to achieve well-defined side lobes that are peaks including the cyclic information signal. Details of the measurement procedure are explained below.

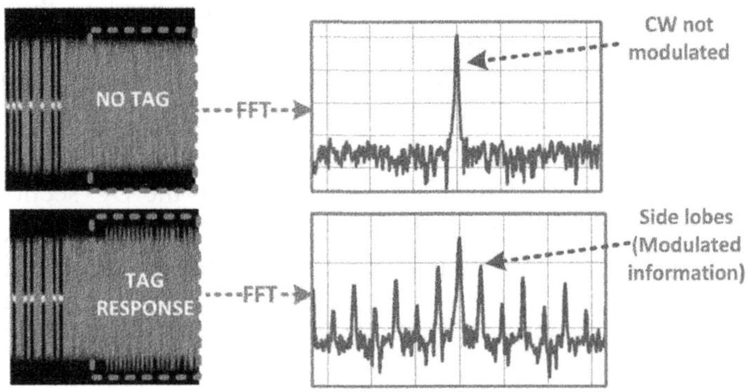

Figure 3.1. *Measurement procedure based on the PSD analysis of UHF RFID signals*

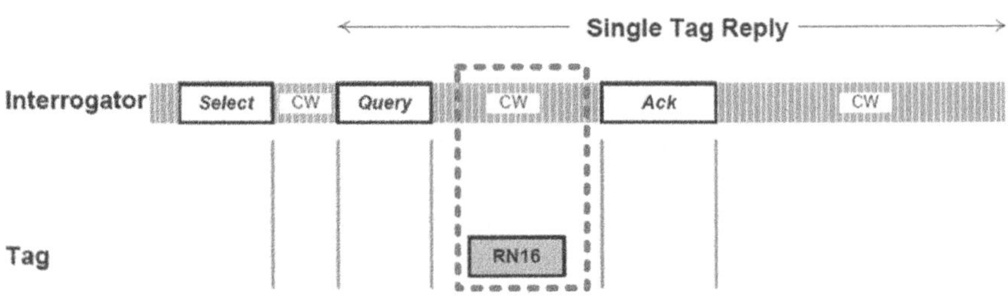

Figure 3.2. *Frame timing in the forward and return links [EPC]*

3.2.3. *RFID air interface*

The standard RFID air interface provides parameter definitions for communication protocols within a common framework for globally accepted frequencies in RFID [ISO 12]. The RFID air interface describes the way in which a tag communicates with a reader by wireless signals and it is part of the physical layer of the multilayer definition of the RFID system presented in Chapter 2.

One of the contributions presented in this chapter is to highlight the characteristics of the backscattering modulation to clearly interpret the time domain signals and their PSD calculation. Characteristics such as backscattering link frequency (BLF), data encoding and preamble are pre-configured on the treatment of the physical layer in the UHF RFID system.

3.2.4. *Configuration of the physical layer in the UHF RFID system*

The configuration of the UHF RFID physical layer is based on the manipulation of the Query command structure sent by the reader. A pre-set Query command orders the target tag to answer with a specific and repeatable PSD, thus allowing clear identification of its spectral content. Two settings play a key role in easing the spectral signal interpretation. The first is to set the tag response (first side lobe from Figure 3.1) far enough from the reader carrier. This is achieved manipulating the BLF.

The second is to force the tag to answer with a periodic signal able to always generate the same spectral shape independently of the tag under test. This periodic signal should last long enough to facilitate the FFT interpretation. This is achieved by configuring the preamble of the 16-bit pseudo random number. And in the same time, the side lobes are well-defined compared with a random sequence.

The tag response to the Query command contains a 16-bit pseudo random number (RN16). The RN16 response has a preamble with a well-known structure depending on how the Query is composed. Figure 3.3 shows the Query structure.

	Command	DR	M	TRext	Sel	Session	Target	Q	CRC-5
# of bits	4	1	2	1	2	2	1	4	5
description	1000	0: DR=8 1: DR=64/3	00: M=1 01: M=2 10: M=4 11: M=8	0: No pilot tone 1: Use pilot tone	00: All 01: All 10: ~SL 11: SL	00: S0 01: S1 10: S2 11: S3	0: *A* 1: *B*	0–15	

Figure 3.3. *Query command structure [EPC]*

The experiences presented in this chapter use the Query 1000101100000000001101. The process to configure suitable values of BLF, the data encoding and the preamble are detailed below.

The first four bits allow us to identify the command as Query command and last five bits correspond to the cyclic redundancy check (CRC5) for error detection and depend on the information content. The information content is the sequence of bits between the Query command identifier and the CRC5. Inside the information sequence, there are three parameters for configuring a suitable tag response.

The bit in the fifth position, which is called the divide ratio (DR), determines where the BLF is defined. The value that the fifth bit can take " 0 " or " 1 " is traduced as DR=64/3 or DR=8, respectively. In the analysis of the PSD, the BLF, which is the spectral distance between the CW component and the tag response, will be higher when the DR is higher. The calculation of the BLF is determined by the ratio $DR/Rtcal$, where $Rtcal=5.5\times Tari$ and $Tari$ is the reference time interval for reader to tag signaling and is equal to the duration of data " 0 " [EPC]. The pair bits in the sixth and seventh positions is called M and sets the tag-to-reader data rate and encoding to which the tag responds. The pair of bits is set to " 01 " which traduces into M=2, which instructs the tag to respond using Miller coding with two cycles per bit as shown in Figure 3.4. Finally, the eighth bit position, which is called TRext, achieves the periodic sequence of bits on the tag response. When M=2 and $TRext$=" 0 ", the tag prefixes a short duration preamble to its RN16 response, and when M=2 and $TRext$= " 1 ", the RN16 response prefixes an extended preamble of 16 bits of data " 0 " with two cycles per bit defined by M=2. Figure 3.5 shows the extended preamble scheme. The periodic sequence has a duration of approximately 65 μs (depending on $Tari$) that is enough to make the PSD calculation. Other parameters on the Query command are not configured since these are used for inventory purposes when the Select command is previously sent which is not the case. However, the values of these remaining parameters do not alter the results.

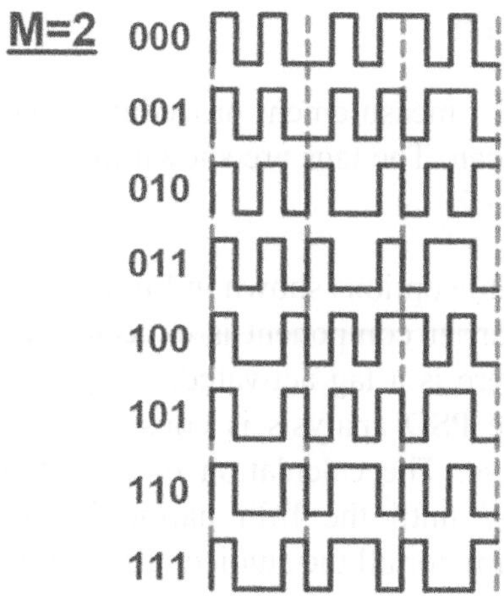

Figure 3.4. *Miller sequences for M = 2 [EPC]*

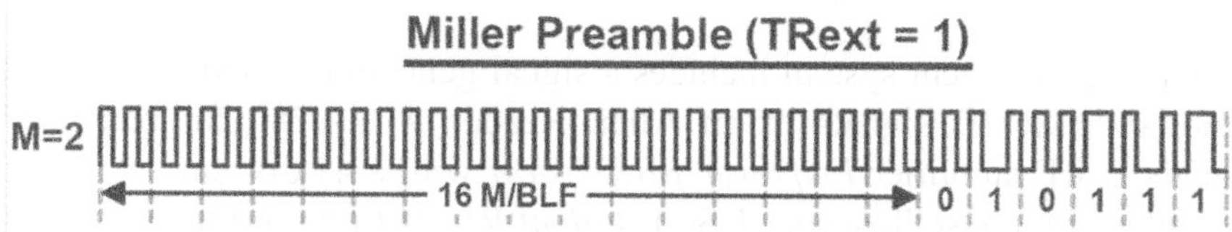

Figure 3.5. *Tag → Reader preamble with Miller M = 2 [EPC]*

3.3. Characterization of RFID tags – radiated measurements

The tag response and harmonic level are evaluated by radiated measurements using commercial RFID inlays. The harmonic level and its relationship with the performance at the fundamental frequency are analyzed in this section. This relation is developed by an experimental parametric study considering the harmonic dependency on the reader power, the bit sequence structure and the channel in the frequency hopping operation. The experimental study is based on calculating the PSD of RFID signals at the harmonic frequencies in the forward and return links in compliance with the EPC Class-1 Gen-2 UHF RFID standard (commercially known as RAIN) [EPC].

3.3.1. *Tags under test*

In order to show the measurement procedure, nine different commercial tags (inlays) were chosen. The tags are shown in Figure 3.7 and their details are provided in Table 3.2.

From the two possible options shown in Figure 3.1, in case of the absence of the tag, only the carrier component is expected in the PSD analysis. On the contrary, when there is a tag activated, the power of the first side lobe around the CW in the PSD analysis is considered to be the power of the modulated tag response. The calculation of the PSD is repeated over the harmonics of the CW until the fifth harmonic order on each tag. The implemented modulation for all the studied tags is the amplitude-shift keying (ASK) [EPC].

3.3.2. *Measurement system*

The measurement system includes a signal generator *Agilent N5182A* for synthesized RFID signals or an Alien Reader ALR-9780 for real RFID signals, a RF amplifier *Empower BBS314AAj* with a 41 dB gain, a 12 GHz Digital Storage Oscilloscope (DSO) *Agilent 91204A* (40 Gsamples/s) and a PC for control and data processing. The set of signal generator and PC is called the RFID tester, this configuration allows managing the Query command. Figure 3.6 shows the measurement setup in the anechoic chamber with a partition wall that provides 48 dB isolation between horn antennas to minimize the mutual coupling:

– The element under test (EUT) tag is located at a distance of 1m from each horn antenna and oriented for maximum reception and re-radiation;

– The forward link signal is generated in only one of two ways: in synthesized signals, the baseband command generated by Matlab is transferred to the signal generator where it is modulated, amplified and sent to the tag via the transmitting antenna; in signals coming from an RFID reader, an ALR-9780 reader is set to replace the signal generator;

– The return link signal is always collected by the receiving antenna, observed and analyzed in the RF band on the DSO;

– In compliance with the EPC Class-1 Gen-2 standard, the synthesized signal carries the Query command. In the case of signals coming from an RFID reader, the RFID air interface cannot be parameterized and standard handshake is made between reader and tag [EPC].

The amplifier was used only in power dependency tests, as will be presented in section 3.7.3.3.

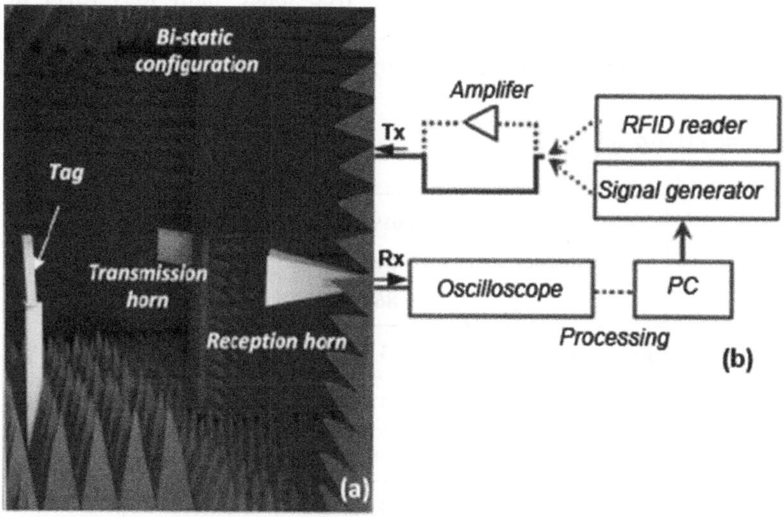

Figure 3.6. *a) Bistatic configuration in the anechoic chamber. b) Equipment setup*

3.3.3. *Power budget*

The power budget in the band 868 MHz–922 MHz considering the measurement setup of Figure 3.6 is described below. The antenna gain A_G varies increasingly between 5.69 dBi and 6.69 dBi at 868 MHz and 922 MHz, respectively:

– in transmission: losses in cables 1, 2 and 3 including connectors are –0.73 dB, –0.73 dB and –1.45 dB, respectively; when the amplifier is used, it inserts a gain of 41 dB;

– in reception: losses in cables 4 and 5 are –0.73 dB and –1.46 dB, respectively;

– the noise level inside the anechoic chamber is approximately –110 dBm.

The power P_{out} delivered by the RFID tester or the reader varies according to the needs of each tag activation. Considering the power budget in transmission and reception and the free-space path loss [3.1] at the distance R of 1m and the wavelength λ at each frequency in Table 3.1, it is possible to calculate each tag sensitivity:

$$L = 20\, log_{10}\left(\frac{\lambda}{4\pi R}\right) \quad [3.1]$$

Frequency	Loss (dB) European band Fc=868 MHz	Loss (dB) US band Fc=915 MHz
Fundamental (Fc)	–31.21	–31.67
Second harmonic	–37.23	–37.69
Third harmonic	–40.75	–41.21
Fourth harmonic	–43.25	–43.71
Fifth harmonic	–45.19	–45.65

Table 3.1. *Free-space path loss*

3.3.4. *Power tag sensitivity*

Power tag sensitivity is defined as the minimum power received by the tag necessary to activate its chip and wait for an understandable response on the reader side. Table 3.2 presents each tag sensitivity at the fundamental frequency calculated as P_{out} minus 28.12 dB in the case of European tags and P_{out} minus 28.58 dB in the case of US tags. These subtraction factors correspond to the described power budget.

Label	Inlay/ Manufacturer	Sensitivity (dBm)	Performance band
T1	Dogbone/UPM [UPM 13a]	–16.89	US
T2	Sec.Frog/UPM [UPM 13b]	–12.93	EU
T3	Web/UPM [UPM 13b]	–11.73	EU
T4	P9-2/LCIS [PER 12]	–11.29	US
T5	MemoryStick/UPM [UPM 13c]	–12.29	US
T6	P10-1/LCIS [PER 12]	–13.63	EU
T7	RapidTrack/Tagsys [RAP 13]	–13.43	EU
T8	Tac.Frog/UPM [UPM 13d]	–13.15	US
T9	AD222 / Avery Dennison [AD]	–11.79	US

Table 3.2. *Details of the evaluated UHF RFID tags*

3.3.5. *Radar cross-section and physical surface of a tag*

3.3.5.1. *Radar cross-section and tag dimensions*

The radar cross-section (RCS) of the tag or echo area is defined as the area intercepting that amount of power which, when isotropically scattered, produces at the receiver a density which is equal to that scattered by the actual object. The RCS σ for dipole antennas of half a wavelength or less, which is the case for common RFID tags, is defined as follows [CON 05]:

$$\sigma = \frac{\lambda^2}{4\pi} G_t^2 \left| \frac{2R_A}{Z_L + Z_A} \right|^2 = \frac{\lambda^2}{4\pi} G_t^2 K \qquad [3.2]$$

where Z_L is the impedance of the antenna load, i.e. the chip impedance, Z_A is the impedance of the tag antenna, G_t is the gain of the tag antenna and R_A is the real part of Z_A. The factor K defines the influence of the chip impedance mismatch on the amount of the RCS. A matched polarization between the reader and tag antennas is considered in [3.2].

It is possible to rewrite [3.2] in terms of the power density of an electromagnetic wave incident on the tag antenna in free space represented by S, and the power backscattered by the tag in the direction of the reader represented by P_{back} as given in [3.3]

$$\sigma = \frac{P_{back}}{S} = \frac{KP_cG_t}{S} \qquad [3.3]$$

where the collected power by the tag antenna P_c is the maximum power that can be delivered to its chip with a complex conjugate impedance. It can be expressed in terms of the effective area of the antenna A_e as [3.4] [CON 05]

$$P_c = SA_e \qquad [3.4]$$

The effective area of the antenna A_e is given by [3.5]

$$A_e = \frac{\lambda^2}{4\pi} G_t \qquad [3.5]$$

Then, [3.3] can be written as

$$\sigma = KA_eG_t \qquad [3.6]$$

The two states of impedance of the RFID chip, which regulate the backscattering modulation technique between the scavenging and reflecting conditions, yield two values of K or consequently two values of the RCS. Then from [3.6], it is possible to define the differential RCS or ΔRCS as

$$\Delta\text{RCS} = \left|\sigma_{scavenging} - \sigma_{reflecting}\right| = |\Delta K| A_e G_t \qquad [3.7]$$

where $|\Delta K|$ means $\left|K_{scavenging} - K_{reflecting}\right|$. The ΔRCS evaluates the functional characteristics of the tag as read range or sensitivity because it is related to the efficiency of the backscattering modulation technique [BOL 10, DOM 10]. Then, finally, the physical area A_p of the antenna can be linked with its effective area by the aperture efficiency ε_{ap} shown in [3.8]

$$\varepsilon_{ap} = \frac{A_e}{A_p} \qquad [3.8]$$

3.3.5.2. *Experimental validation*

From [3.6] and [3.8], by considering the physical area of the tag antenna for a given tag, the larger the resonance surface, the larger the effective area and the higher the RCS. Figure 3.7 shows 20 μs of tag response for the nine tags where the modulated backscatter is clearly distinguishable. Tag T1, which exhibits a larger resonance surface, has a higher scatter capability than the RCS. In an ideal condition when K=1 in the scavenging state, an antenna loaded with a complex-conjugate impedance load scatters the same amount of power as absorbed by the load, thus causing a maximum ΔRCS. Although all tags are illuminated with the same power P_{out} = 16.75 dBm, it is highlighted in Figure 3.7 that the amplitudes, directly related to the ΔRCS, vary up to 42%. P_{out}=16.75 dBm represents the power to activate the tag with less sensitivity T4.

Consequently, there are two factors involved in the $|\Delta RCS|$ variations [3.7]:

– the matching degree between the antenna and the RFID chip in the scavenging state;

– the physical area of the antenna.

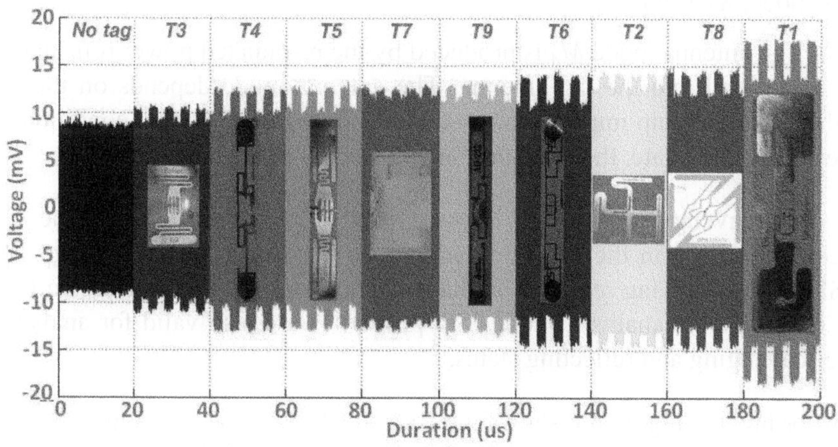

Figure 3.7. *Time domain response from nine UHF RFID tags. For a color version of this figure, see www.iste.co.uk/andia/non-linearites.zip*

3.3.6. *Optimized PSD analysis of the RFID communication*

3.3.6.1. *PSD properties*

The interpretation of the PSD is enhanced due to the contribution of the RFID physical layer setting which achieves a periodic sequence of bits on the tag response. This implies that the characteristics of the modulated signal coming from the tag are enhanced on the observed window. Indeed, in this case, the amplitudes of the spectrum are maximal and the BLF is distinguishably positioned at 230 kHz from the CW. It should be noted that for the following analysis, the power of the tag response will be quantified as the amplitude of first side lobe. Figures 3.8–3.10 show examples of the PSD calculation while the communication of the reader to the tag is established. The PSD at CW fundamental and harmonic frequencies are compared with the PSD when no tag is present. In this last case, the observed CW is essentially due to the coupling between horns.

Two operational modes of the passive UHF RFID tag are defined in both states of modulation:

– The structural mode M_S is produced when the tag antenna is in the matching condition. The scattered power contribution is always constant. In this mode, K=1 in [3.2];

– The antenna mode M_A is produced by the re-radiated power reflected by the chip load into the tag antenna. The antenna mode depends on the two values that the chip impedance can take. Under ideal matching conditions in the scavenging state, the structural mode is equal to the antenna mode.

The scavenging state can be quantified at the PSD level at the CW frequency. Only in the reflecting state does the antenna mode change. The PSD level of the tag response in the reflecting state is observed at the BLF. In the harmonic analysis, these PSD properties are also valid for analyzing the scavenging and reflecting states.

For tag T5, Figure 3.8 shows the PSD at 915 MHz while Figures 3.9 and 3.10 show the PSD at the second harmonic (i.e. 1830 MHz) and the third harmonic (i.e. 2745 MHz). The tag T5 was chosen because it has a higher PSD level at the third backscattered harmonic. In Figures 3.8 and 3.10, the tag response with modulated information on the side lobes like a typical spectrum of the square waves is clearly seen. In the case of the second harmonic shown in Figure 3.9, no response was observed at the activation power.

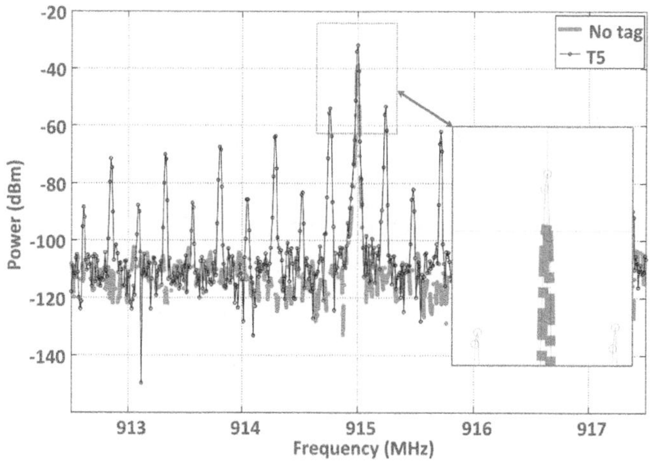

Figure 3.8. *PSD at the fundamental frequency for tag T5 when P_{out} =15.6 dBm. For a color version of this figure, see www.iste.co.uk/andia/non-linearites.zip*

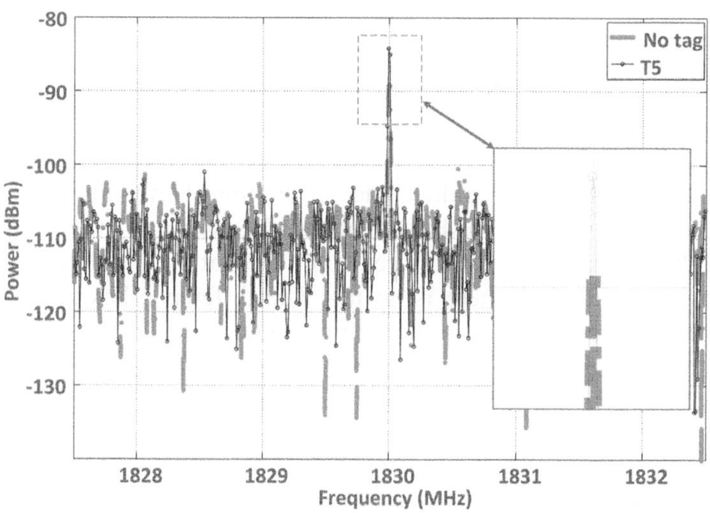

Figure 3.9. *PSD at the second harmonic frequency for tag T5 when P_{out} =15.6 dBm. For a color version of this figure, see www.iste.co.uk/andia/non-linearites.zip*

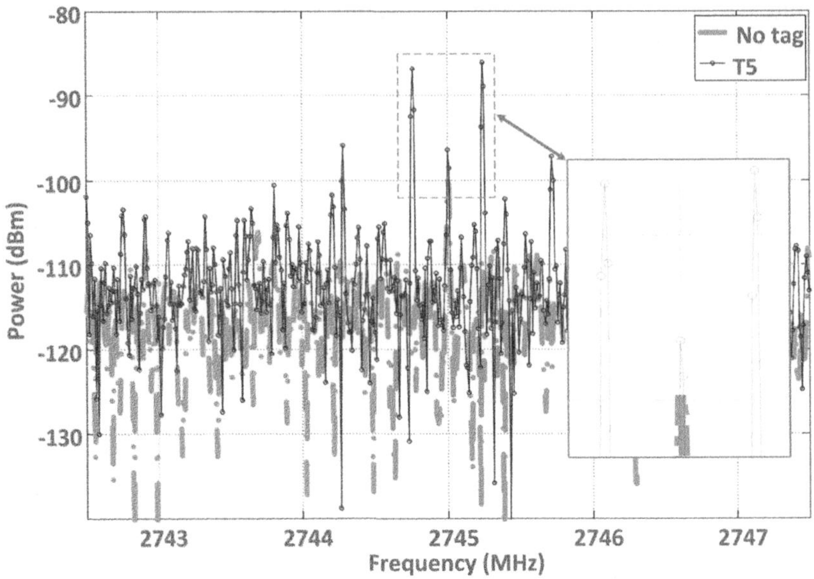

Figure 3.10. *PSD at the third harmonic frequency for tag T5 when P_{out} =15.6 dBm. For a color version of this figure, see www.iste.co.uk/andia/non-linearites.zip*

3.3.6.2. *Matching in the scavenging state*

As shown in Figure 3.8, when there is a tag, the PSD level of the CW at the fundamental frequency in the presence of a tag is 6.9 dB higher compared with when there is no tag. The 6.9 dB are taken as the total contribution of M_S+M_A. The difference at the third harmonic is 3.27 dB, as shown in Figure 3.10. In the PSD analysis, the matching in the scavenging state can be quantified by the difference of levels at the CW frequency when there is a tag and when there is no tag. The difference has to be minimal representing only the RCS in structural mode. Large differences indicate mismatch reflections. The difference indirectly assesses the characteristics of the RCS from each tag at fundamental and harmonic frequencies.

Figure 3.11 shows the PSD value of the CW fundamental frequency and CW harmonics for five representative tags until the fifth harmonic, considering each tag sensibility. The mesh represents the measured CW power when no tag is present also until the fifth harmonic. A decreasing behavior is observed while the harmonic order increases. The only exception is tag T1, which has a second backscattered harmonic weaker than the third, revealing scavenging properties at the second harmonic. Tag T5 has the best scavenging performance, i.e. the better matching at the third harmonic according to the criterion of minimal difference of PSD levels at the CW frequency.

In the performed setup, the PSD calculation at the CW frequency represents the coupling between the horn antennas, which reveals CW harmonics coming from the RFID tester. Scatter capabilities of each tag vary the level of the observed CW, which depends on the RCS of each tag.

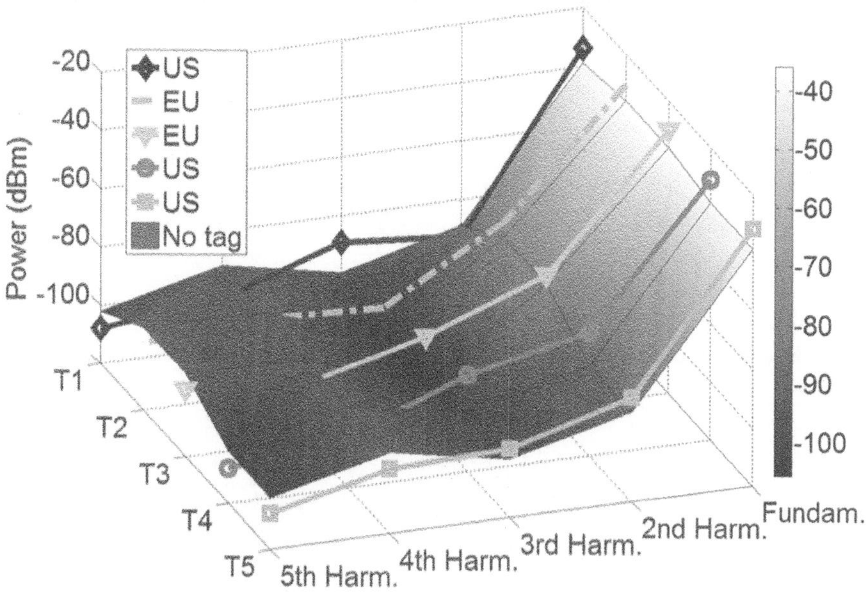

Figure 3.11. *Measured PSD at CW harmonic frequencies with P_{out} considering each tag sensibility. For a color version of this figure, see www.iste.co.uk/andia/non-linearites.zip*

3.3.6.3. *Harmonic backscattering*

The level at the BLF in the PSD analysis at each CW harmonic represents the backscattered harmonics in the reflecting state. It is important to explain the basis of harmonic radiation of the tag antenna, usually designed as an optimization process of a half-wave dipole at the fundamental frequency. Figure 3.12 shows the harmonic current distributions on a half-wave dipole operating at the fundamental frequency, which shows that for even harmonics the current at the antenna input is zero. Although if even harmonics were produced by the RFID chip, they would not be efficiently re-radiated since the half-wave dipole would require infinite impedance at even harmonics, which is practically impossible [CHA 89]. Figure 3.13 shows the level contributed by five representative tags on their modulated response at each harmonic frequency. In all cases, the fundamental and third harmonic responses are predominant. Even though the fifth is an odd harmonic, it has a burly distinguishable level. Then, the claim of a perceptible third backscattered harmonic carrying information is valid.

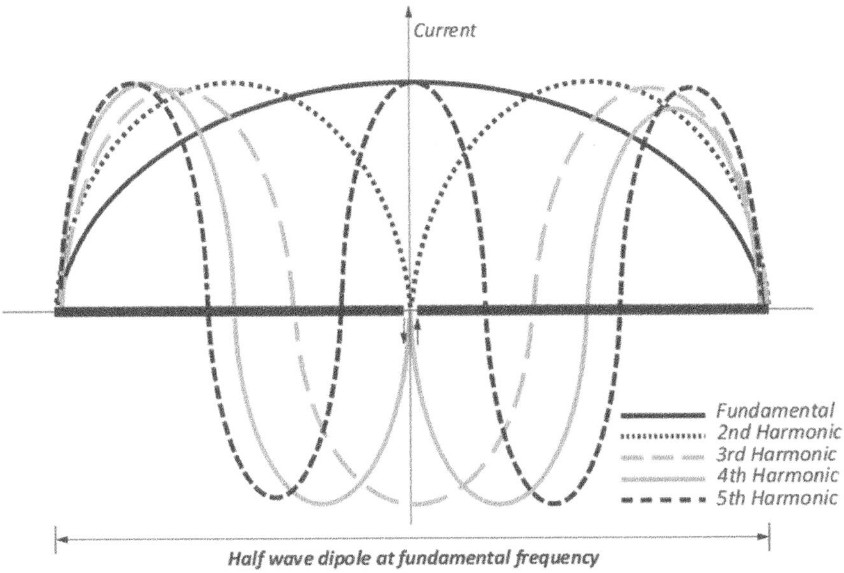

Figure 3.12. *Current distribution until the fifth harmonic along the length of a half-wave dipole designed for the fundamental frequency. The half-wave dipole requires zero admittance at even harmonics to be able to radiate*

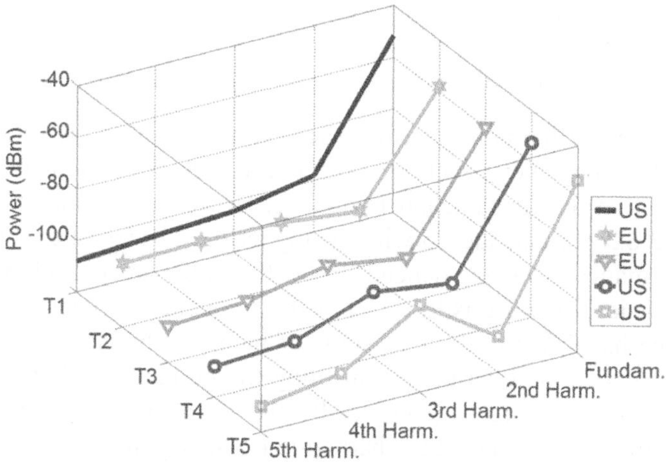

Figure 3.13. *PSD of the tag response at the BLF. Pout considers each tag's sensibility*

The higher power of the first side lobe on the third backscattered harmonic is for tag T5 with –86.03 dBm. No significant second harmonic was detected, because its measured value was below –86.03 dBm. Table 3.3 summarizes the results for the second and third harmonics for all tags regarding each tag's sensitivity.

Tag	Second harmonic Power (dBm)	Third harmonic Power (dBm)
T1	–98.11	–104.1
T2	–98.29	–95.08
T3	–102.9	–98
T4	–99	–94.72
T5	–106	–86.03
T6	–100	–92.08
T7	–104.7	–104.5
T8	–102.4	–98.24
T9	–98.39	–93.34

Table 3.3. *Backscattered harmonics from the tag response*

3.3.7. *Dependency analysis of harmonic scattering*

3.3.7.1. *Dependency on operation at the fundamental frequency*

Analyzing the harmonic performance with respect to the results at the fundamental frequency, section 3.3.5.1 presents tag T1 with the best performance at the fundamental frequency. However, the analysis made in section 3.3.6.3 presents tag T5 with the higher third backscattered harmonic. The performance of the tag depends basically on the antenna–chip matching and its physical characteristics [3.7] at the frequency of interest. Since the tag antenna has different designs, some of them can be close to a matching situation on some odd harmonics profiting the current distribution over the dipole shown in Figure 3.12, especially the third one due to its level. A certain degree of matching at the harmonic frequency may cause the antenna to re-radiate the harmonics produced by the RFID chip. However, the harmonic generation capability is defined by the chip itself. In order to quantify the effect of the tag antenna impedance on the level of radiated harmonics, the characterization must be performed directly in the chip. This topic will be addressed in section 3.4.

3.3.7.2. *Dependency on bit sequence*

Achieving the manipulation of the RFID air interface to reproduce tag responses with different bit structures enables the development of a bit sequence dependency analysis. The dependency on the sequence of bits on the backscattered harmonic is studied by calculating the PSD over a random bit sequence, by randomly choosing a portion of the RN16 response instead of the preamble. Therefore, the comparison of PSD was made on two different sequences when P_{out}=16.75 dBm. Figure 3.14 shows the PSD comparison at the fundamental frequency for tag T4. As expected, the side lobes in the periodic sequences are narrow and powerful compared to those in the random sequences. The same behavior is observed for the third harmonic in Figure 3.15. In order to compare the power levels of the modulated tag response until the fifth harmonic from the periodic and random sequences, Figure 3.16 shows the data for tag T5. The PSD level of periodic signals is higher until the fourth harmonic. In the fifth harmonic, the relation is inverse, but close to the noise level. Then, the sequence of bits transmitted by the tag directly influences the backscattered harmonics. The merit of the air interface configuration is that it can reproduce a tag response carrying a periodic sequence of bits over all EPC Class-1 Gen-2 tags facilitating the PSD analysis. Looking at the applications, the use of periodic sequences can be advantageous when the detection of signals from different sources and diverse spatial antenna arrangements produces an interfered spectrum in which the detection of the desired signal is difficult.

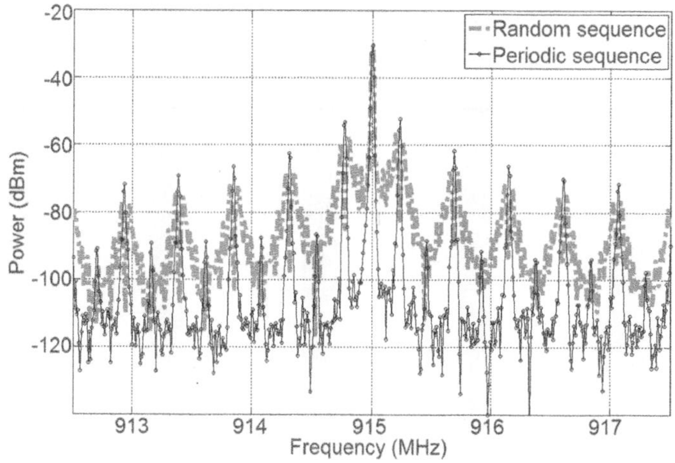

Figure 3.14. *PSD of the periodic and random sequences of bits in the response of tag T4 at the fundamental frequency. For a color version of this figure, see www.iste.co.uk/andia/non-linearites.zip*

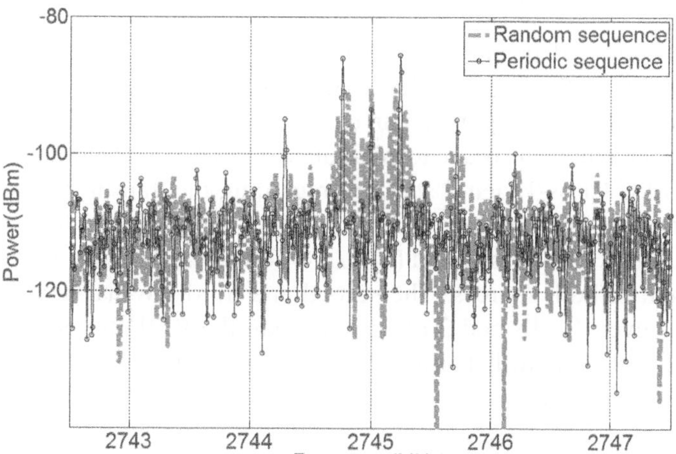

Figure 3.15. *PSD of the periodic and random sequences of bits in the response of tag T5 at the third harmonic. For a color version of this figure, see www.iste.co.uk/andia/non-linearites.zip*

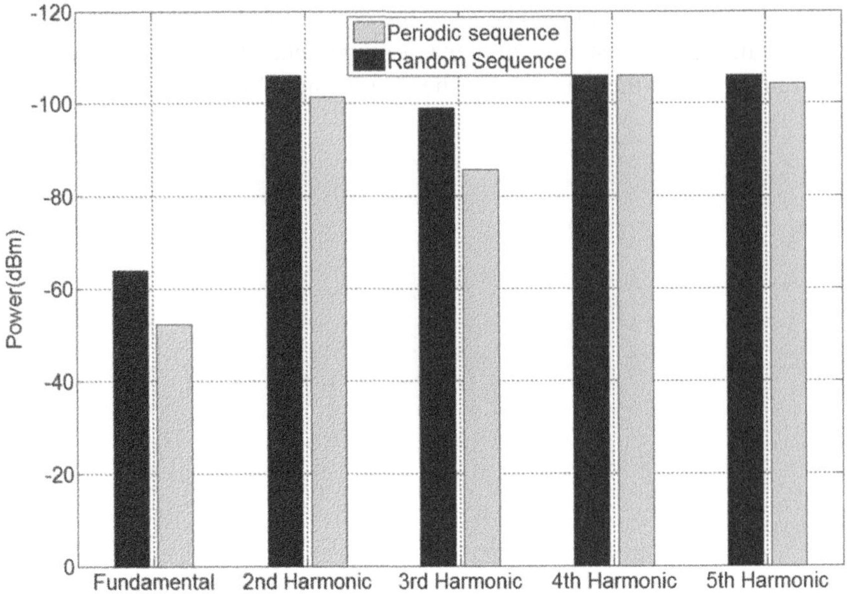

Figure 3.16. *Comparison of PSD levels from the periodic and random sequence responses for tag T5*

3.3.7.3. *Dependency on reader power*

Reader power dependency is studied by varying P_{out} from the minimum level to activate the tag with less sensibility until reaching the equipment limit or to saturate the tag in certain cases. No low-pass filter (LPF) was used at the reader output in order to reproduce a normal scenery in compliance with the EPC Class-1 Gen-2 standard regarding the transmitted power level at fundamental and spurious bands. For this experience two sources were considered: (1) RFID tester and (2) *ALR-9780 reader*. Figure 3.17 shows the comparison of both sources. The second and third CW harmonics from the reader are even 14 dB and 2 dB greater than the RFID tester, respectively. For the fourth and fifth harmonics, the reader and the RFID tester have comparable power levels.

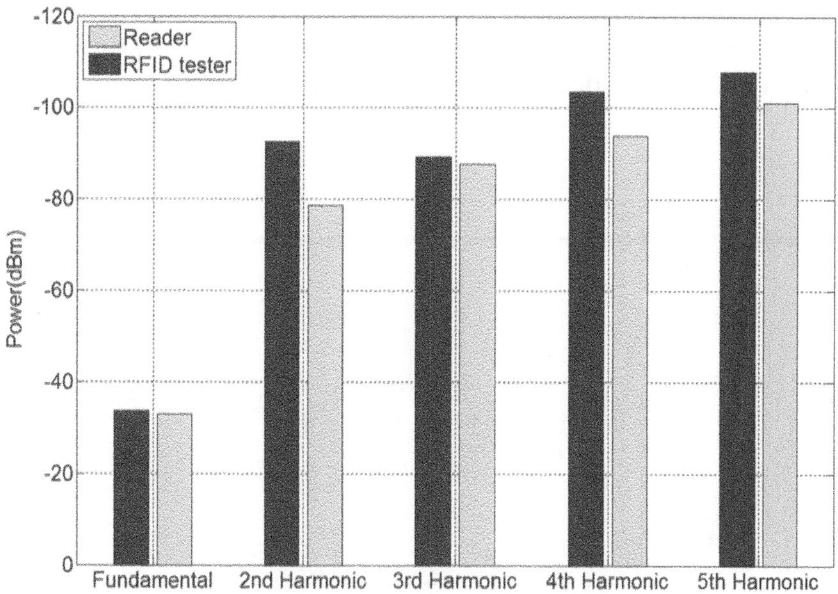

Figure 3.17. *Comparison of harmonic levels generated from the RFID tester and one commercial reader*

A second set of tests was performed by introducing the amplifier after the signal generator, which allows evaluation at larger power ranges. The amplifier introduces 41 dB of gain. P_{out} is first set to –24.2 dBm, which represents an average power of operation sufficient to work with all available tags. An increment of 19.2 dB which means P_{out} = –5 dBm does not affect the strength of the tag response at the fundamental frequency as can be seen in Figure 3.18 but only affects the level at the CW frequency. On the contrary, Figures 3.19 and 3.20 show that this effect is clearly constructive for the second and third harmonics of the tag response. Then, the backscattered harmonics directly depend on the power level that the reader transmits. Two reasons can explain this effect both based on the increment of the harmonic level from the reader:

– By increasing the reader power, the input impedance of the chip changes because of its non-linear behavior and this changes the matching with the antenna. Depending on this, the shunt transistor, which regulates the internal supply voltage of the digital part, receives power to activate it and works until its upper power limit [LUK 07];

– The tag could respond to the harmonic Query. The effect in which the tag responds at various frequencies is called *backscattering frequency diversity* and was reported in [MIN 11, HIS 07].

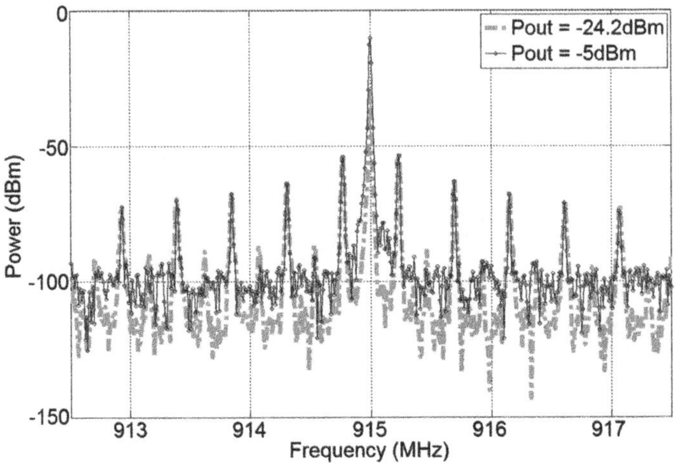

Figure 3.18. *Effect of the transmitted power on the backscattered response at the fundamental frequency. For a color version of this figure, see www.iste.co.uk/andia/non-linearites.zip*

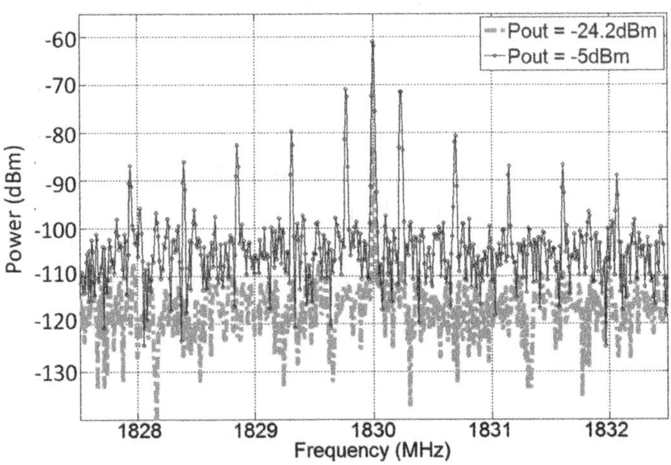

Figure 3.19. *Effect of the transmitted power on the second backscattered harmonic. For a color version of this figure, see www.iste.co.uk/andia/non-linearites.zip*

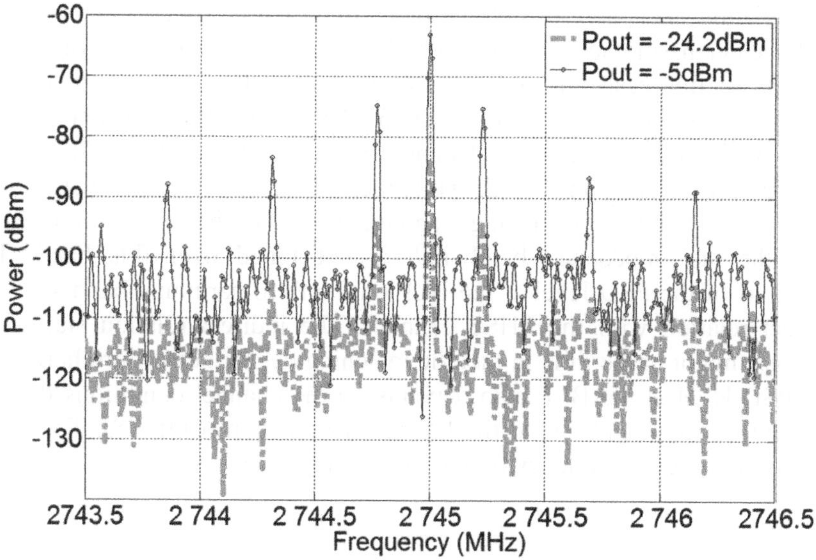

Figure 3.20. *Effect of the transmitted power on the third backscattered harmonic. For a color version of this figure, see www.iste.co.uk/andia/non-linearites.zip*

3.3.7.4. *Dependency on frequency hopping*

The frequency hopping spread spectrum (FHSS) technique is based on the automatic hopping of the CW frequency at brief time intervals. The frequency is chosen in a pseudo random way from a set of frequencies covering a much wider band than the bandwidth required to transmit the signal containing the useful data. In the US, according to the current Federal Communications Commission (FCC) regulation, the FHSS can be used to sweep the frequency band from 902 MHz to 928 MHz with 500 kHz width per channel [EPC]. Figure 3.21 shows the PSD level at CW fundamental frequencies over the entire US band for tags T1, T4 and T5. The top curves represent the PSD at the CW frequency and bottom curves represent the PSD level at the BLF. The CW level varies 3dB over the entire band. However, the bottom curves are almost constant. As was described before, the impedance of the RFID chip varies according to the received power. Then, this variation is capable of helping or harming the chip performance at the fundamental harmonic depending on the impedance bandwidth of the tag antenna. In this case, the 3 dB variations on the CW level do not affect the tag response in the US band. Figure 3.22 shows the PSD level at the third harmonic for the same tags T1, T4 and T5. The PSD level at the BLF on the third harmonic varies until 10dB while the PSD level at the BLF varies 5dB. The non-constant level of the tag response at the third harmonic is consistent since the tag impedance bandwidth is not tuned over this band. The variations of the PSD level at the BLF on the third harmonic band show an indirect relation with the FHSS channel. The indirect relation states that there is a major cause: it is in fact the matching degree at the third harmonic. The PSD level at the BLF also depends on the channel conditions that are different in each band due to the fading phenomenon [COL 98]. Therefore, there is no direct dependency on the FHSS technique.

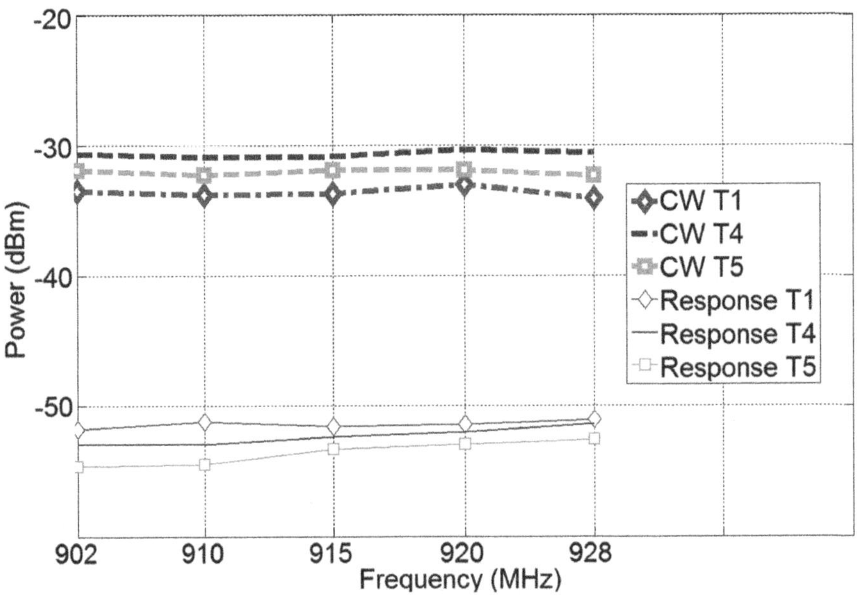

Figure 3.21. *Power on frequency hopping channels at the fundamental frequency*

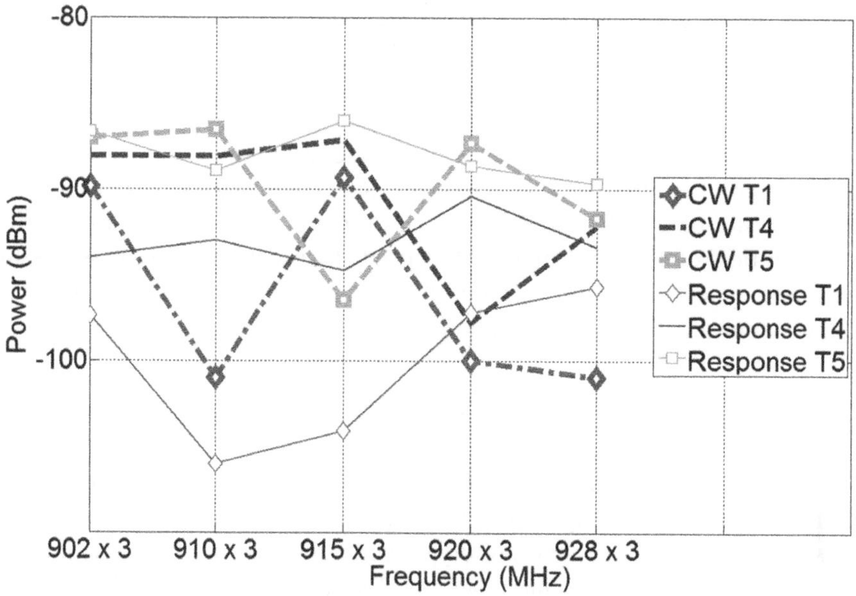

Figure 3.22. *Power on frequency hopping channels at the third harmonic*

3.3.8. *Limitations of tag characterization by radiated measurements*

A practical experiment to measure the non-linearities in the RFID UHF system was described throughout section 3.3. The measurement procedure is based on the PSD analysis performing a temporal acquisition directly in the RF band while the communication from the tag to the reader is fully established. By setting the Query parameters, it is possible to manipulate the RFID physical layer in order to facilitate the PSD interpretation. The PSD calculation reveals harmonics on the return signal consequence of the not treated non-linearity effects of the RFID chip. It was also noted that harmonics are generated by the reader in the forward signal. The uncontrolled non-linearity in the current UHF RFID technology allows the tag to generate harmonic signals carrying information without constraints to the EPC Global Class-1 Gen-2 standard. The backscattered harmonics are independent of the FHSS transmission technique but depend on the power sent by the reader and the sequence of bits that the tag transmits. To investigate the effect of the tag antenna at the fundamental frequency on the level of backscattered harmonics, a deeper analysis at chip level is necessary.

The analysis made on this section considers the entire tag (antenna and RFID chip) as the EUT. The analysis evaluates the non-linearity characteristic of a UHF RFID EPC Class 1 Gen 2 tag without any specific consideration of the kind of the chip used. In order to directly assess the harmonic generation from the chip, the next section will focus on the heart of the matter, i.e. the characterization of the RFID chip.

3.4. Characterization of RFID chips–conducted measurements

In passive UHF RFID systems, the challenge is always to develop reduced size and cheap tags. For this purpose, new test platforms, characterization methods and prototype techniques are definitely required to accurately characterize the tag performance and propose design improvements. The existing test platforms, such as in [SUN 09, SUN 10, CAT 12, DED 13, CAT 11, NIK 09a, NIK 09b] are based on the measurement of, at that time, traditional features of the tag, e.g. the chip-antenna integration [SUN 09], impedance matching condition [SUN 10, CAT 11, NIK 09a, NIK 09b] or differential radar cross-section (ΔRCS) and sensitivity [CAT 12, DED 13, CAT 11, NIK 09a]. In addition, most of the methods to characterize the RFID chip are performed without activating the chip during measurements [SUN 09, SUN 10, NIK 09a, NIK 09b] and always at the fundamental frequency.

Commercial test benches such as Tagformance from Voyantic [TAG] and RFID Xplorer from CISC Semiconductors [RFI] are the plug and play solutions addressed primarily to industrial RFID actors who are looking for tag design optimization using the tag activation threshold as the key performance indicator. Depending on the test requirements and parameters for evaluation or optimization, limited by standard functionalities, one of the reported laboratories or commercial test benches solutions could be suitable. However, when exploring for new functionalities of standard RFID tags and measuring advanced parameters, especially those caused by their non-linear behavior, existing test platforms and methods are not suitable.

Previous sections showed the existence of modulated backscattering signals backscattered by the tag at the harmonic frequencies of the CW. So far, it is known that the diode-based rectifier operation performed by the RFID chip determines the non-linear behavior of the chip resulting in harmonic signal production [DEV 05, BOL 10] where the impedance modulation is still present, producing a redundant backscattering modulation. This concept is presented in Figure 3.23, showing an RFID tag backscattering information not only at the fundamental frequency (e.g. 868 MHz) but also, for example, at the third harmonic (i.e. 2604 MHz). Further details about the rectifier architecture and operation will be discussed in Chapter 4.

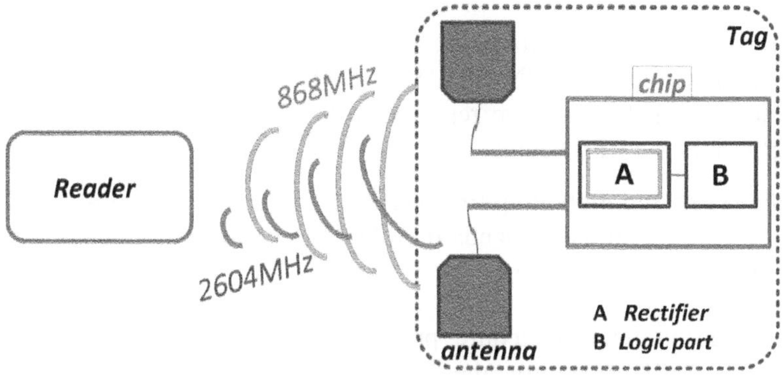

Figure 3.23. *Illustration of the presence of multiple channels in the tag-to-reader communication link. The impedance modulation is still present in the harmonic signal*

As was experimentally demonstrated in previous sections by means of radiating measurements, commercial tags radiate some of the reflected harmonic signals generated by the rectifier, due to the independent design of the antenna and the RFID chip [RAO 05]. In order to delimit and characterize the source of non-linearities in RFID tags, the definition of a specific test platform, which enables accurate measurements during RFID chip operation in a wide frequency range (including harmonics), is needed. With this in mind, this section aims to meet two main objectives:

– to define a new RFID non-linear test platform (RFID-NTP) to perform a thorough characterization and a study of the non-linear phenomena (particularly harmonic signals carrying information during an RFID communication);

– to carry out an experimental study to show the effect of the non-linear treatment from the tag antenna side (basically based on antenna–chip matching experiences at different bands).

3.4.1. *Non-linear characterization platform*

Regarding the feature functionalities of the above-reported characterization platforms, three groups of contributions are highlighted below as a guide to developing the envisaged RFID-TP:

– so far, all contributions have reported measurement techniques and methods only at the fundamental frequency;

– most of the contributions do not allow activation of the RFID chip while characterizing it [SUN 09, SUN 10, NIK 09a, NIK 09b], with the only exception being the platform reported in [LUK 07], which uses a special fixture to hold the RFID chip;

– the work reported in [NIK 09a] and [NIK 09b] proposes a simple fixture, by means of using standard 50 Ω SMA connectors to hold the RFID chip without special matching.

The measurement procedure presented in this section combines two techniques: [NIK 09a] for the simple chip fixing and [LUK 07] for measurement in temporal and frequency domains while the chip is activated. In addition, the measurement is performed in a wide frequency range that allows a characterization until the fourth harmonic (i.e. 868 MHz). The waking up of the chip, according to the EPC Class-1 Generation-2 UHF RFID standard [EPC], was done as in the measurement setup described in section 3.3. The analysis method is based on the measurement of the PSD. Unlike the radiated measurements presented in section 3.3, the RFID-NTP includes a vectorial characterization of the RFID chip. Consequently, one of the contributions of the RFID-NTP is the ability to measure both impedance modulation states (scavenging and reflecting) in a wide frequency range.

3.4.2. *System operation description*

The block diagram of the RFID-NTP is shown in Figure 3.24. Two main parts are highlighted: the first one consists of specialized equipment for microwave measurements and the so-called RFID tester; the second one is based on two impedance tuners connected in series and provides the complex conjugate of the chip impedance of the scavenging state which, in fact, is the optimum antenna impedance. The last part is the so-called antenna impedance emulator. With the described configuration, a complete characterization can be performed under full communication between RFID tester and chip. The setup of the RFID-NTP shown in Figure 3.25 is described below.

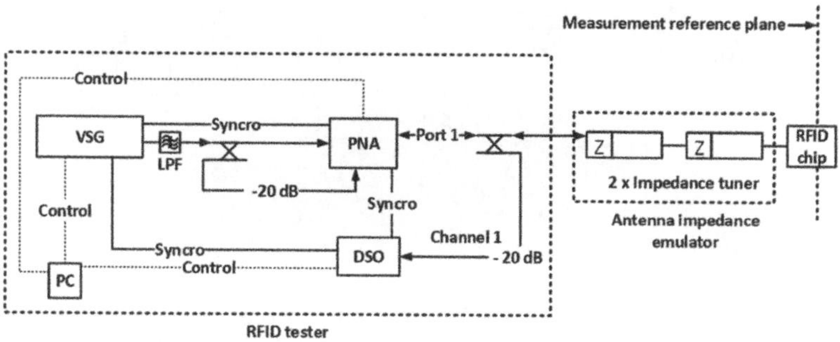

Figure 3.24. *Structure of the RFID-NTP used to characterize the backscattered harmonics by RFID chips*

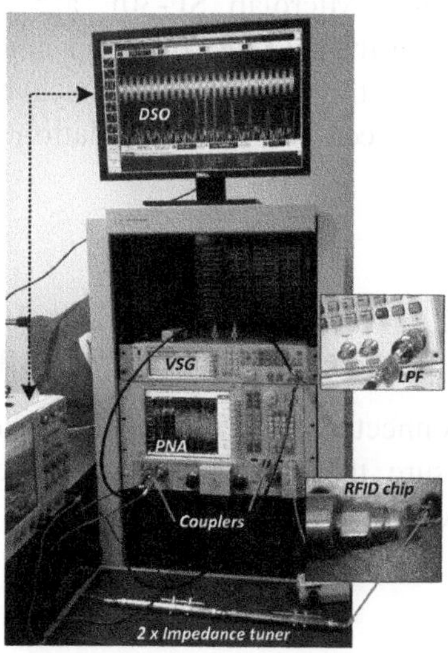

Figure 3.25. *RFID-NTP with the connected RFID chip*

3.4.2.1. *RFID chip fixture*

RFID chips were fixed over 50 Ω SMA connectors [NIK 09a]. Soldered connections and gluing using conductive ink connections are possible. The chip fixtures and calibration kit are shown in Figure 3.26(a).

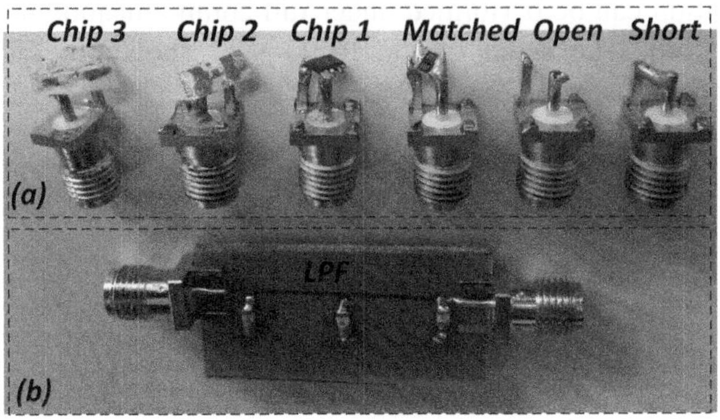

Figure 3.26. *a) RFID chips and calibration kit. b) LPF used to reject harmonics from the RFID tester*

3.4.2.2. *Antenna impedance emulator*

Two impedance tuners Microlab SF-30F are connected in series to provide matching between the 50 Ω RFID tester and the RFID chip. When the matching is achieved, the chip is in the scavenging state and, once activated, it switches between the two modulation states. The details are shown in [LUK 07].

3.4.2.3. *RFID tester*

– An Agilent N5182A Vector Signal Generator (VSG) is used to generate and send the Query command (1000) [EPC] at 868 MHz.

– A 20 dB LPF is connected at the RF output of the VSG in order to filter its harmonics and ensure the chip harmonic response is self-generated. Figure 4.4(b) shows the adopted LPF.

– An Agilent N5224A Performance Network Analyzer (PNA) is used as an external source in the VSG. After the Query command is sent, the PNA determines the impedance of the chip during the scavenging and reflecting states over a temporal sweep at the frequencies of interest. The use of the PNA allows performing a vector characterization of the RFID chip.

– A first C123E-20 ATM 20 dB directional coupler is used in order to inject the signal coming from the VSG (input port of the coupler) to the PNA (transmitted and coupled ports of the coupler). This setup allows the proper configuration of the external source for the PNA [AGI 06].

– A second C123E-20 ATM 20 dB directional coupler allows visualizing the chip response from the coupled port, on the oscilloscope, and to separate it from the CW. The visualization allows setting the optimal position of the impedance tuners by minimizing the CW level. The direct port goes to the PNA.

– A 12 GHz DSO Agilent 91204A (40 Gsamples/s) connected to the coupled port of the second coupler allows us to clearly observe the chip response. In addition, a PSD analysis is performed over the time window to assess the level response of the chip.

– Output power, timing, synchronization, triggering and data acquisition are controlled by a Matlab program running on the host PC.

3.4.2.4. *Calibration*

The flow diagram shown in Figure 3.27 describes the calibration and measurement procedure used for each chip. The minimum activation power for a given chip was found by an iterative up-and-down process until the PSD detected a tag response. In the calibration step, a first traditional calibration at the measurement reference plane shown in Figure 3.24 was done using the PNA E-cal module [KEY 14]. After this, the de-embedding of the SMA connector was performed by means of a second calibration at the same plane, now using the pre-defined short-open-load (SOL) kit shown in Figure 3.26(a). In the power budget calibration, the losses of the directional couplers, cable and connector must be considered. Finally, the characterization of harmonic responses consists of applying the same procedure, but now by setting the temporal sweep of the PNA and the PSD analysis in the DSO to the frequency of the harmonic. Note that the VSG only transmits at the fundamental frequency of 868 MHz and the chip activation is always observed at this frequency.

This process was repeated for each chip under test. The data acquisition was automated by a Matlab script which uses the Agilent Command Expert tool [TEC] to easily program the interconnection between equipment, scope options and sweep parameters.

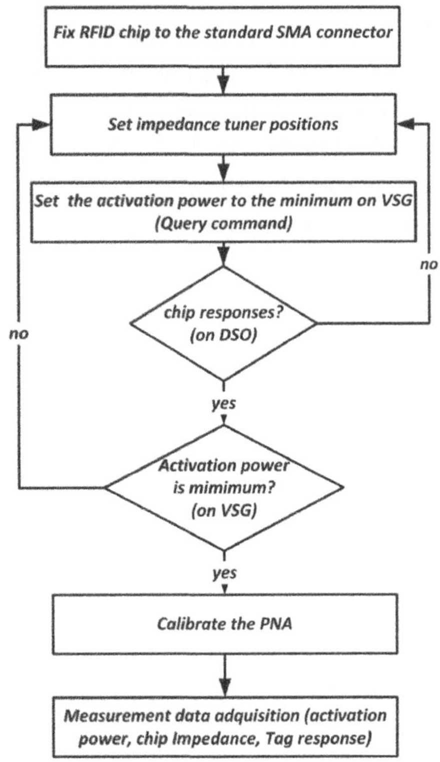

Figure 3.27. *Flow diagram of the measurement procedure in the RFID-NTP*

3.4.3. *Activation threshold and impedance measurement*

The RFID-NTP described in section 3.4 was used to measure and characterize several RFID chips. As an example, this section presents the measurement results for three commercial RFID chips: (1) G2XM NXP SOT505-1 [NXP 13], (2) G2XM NXP SOT1040-AA1 [NXP 13] with strap and (3) Impinj Monza 5 [IMP] extracted from the commercial inlay AKtag [TAG]. The results of the PSD analysis in the DSO, allowing the observation of the chip response in a frame, are shown in Figure 3.28.

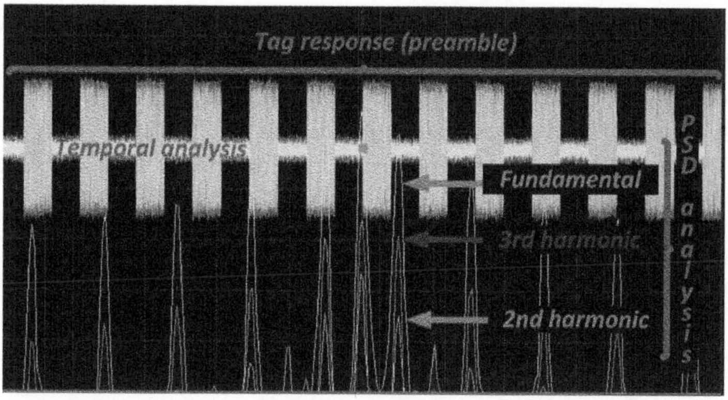

Figure 3.28. *Harmonic characterization method in the DSO. The visualization allows the optimal position of the impedance tuners to be set by minimizing the CW level*

A power sweep was performed to find the chip activation power while keeping the impedance tuners fixed at the position found during calibration. For each power step, the response level and chip input impedance for the fundamental and its harmonics up to the fourth were measured. Figure 3.29 shows the response level at the fundamental frequency for the three chips. The waking up point for each chip determines its minimum operating power, also known as chip sensitivity. For these three chips, the sensitivity varies between –13.2 dBm and –11.5 dBm. A chip sensitivity of −15 dBm, measured at the chip pads, is reported in datasheets [NXP 13]. The difference from the results presented here is due to the losses on the test fixture, considering that on-chip measurements have slightly higher accuracy compared to this method. In the case of chip 3, a saturation point at higher power levels is observed. This is due to the shunt control that regulates the internal supply voltage for the digital part of the chip. The saturation is above 8.8 dBm for this chip.

Figure 3.30 shows the measured impedances during the scavenging state for chips 1 and 3 at the fundamental frequency. Impedance results are in good agreement with values reported in the datasheets (see Table 3.4) at the activation power. Although chips 1 and 2 have the same IC component, they have different packages, which explains the difference its impedance values. The optimum antenna impedance is the complex conjugate of the chip impedance at the sensitivity point shown in Figure 3.29. Quite significant changes in the impedance can be observed while increasing the input power. These changes show the non-linear behavior of the chip, more precisely due to the rectifier characteristics [DEV 05, ROB 12]. A relatively constant impedance can be observed at low input power levels, once the chip is activated. This is explained by the fact that, as electromagnetic energy harvesters, RFID chips are also optimized to start operating at low input power. The non-linear effect depends on the number of rectifier stages used by each RFID chip [BOA 13].

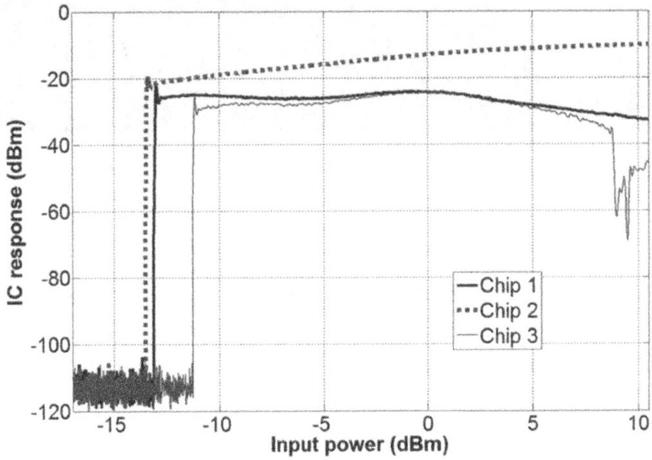

Figure 3.29. *Measured power sensitivity of RFID chips*

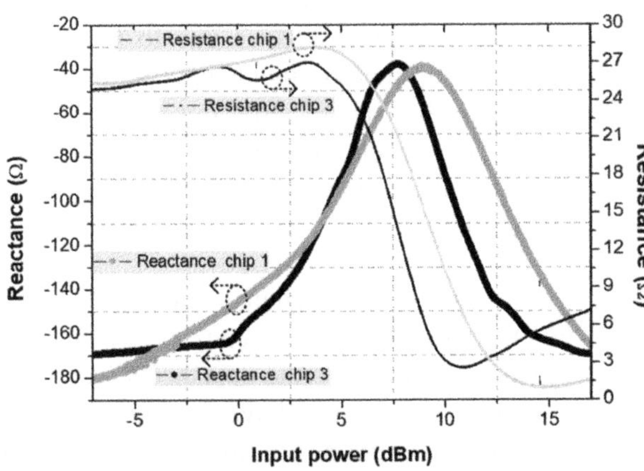

Figure 3.30. *Impedance for chips 1 and 3 in the scavenging state for a sweep of power*

Chip	Measured (Ω)	Datasheets (Ω) [NXP 13, IMP 16, IMP 17]
Chip 1	22 - j182	22 - j195
Chip 2	19 - j178	22 - j195
Chip 3	20 - j170	16 - j171

Table 3.4. *Impedance results at the fundamental frequency*

3.4.4. *Harmonic characterization*

The harmonic response levels were measured for the three chips until the fourth harmonic by performing the PSD analysis at the same frame response. The results are shown in Figure 3.31. As already discussed in section 3.3 and confirmed here, the third harmonic is predominant for all the chips under test. It is worth noting that, even until the fourth harmonic, a modulated response is detected. This becomes clearer for an input power above 3 dBm in the case of chips 1 and 3. The details of the activation power are shown in Table 3.5.

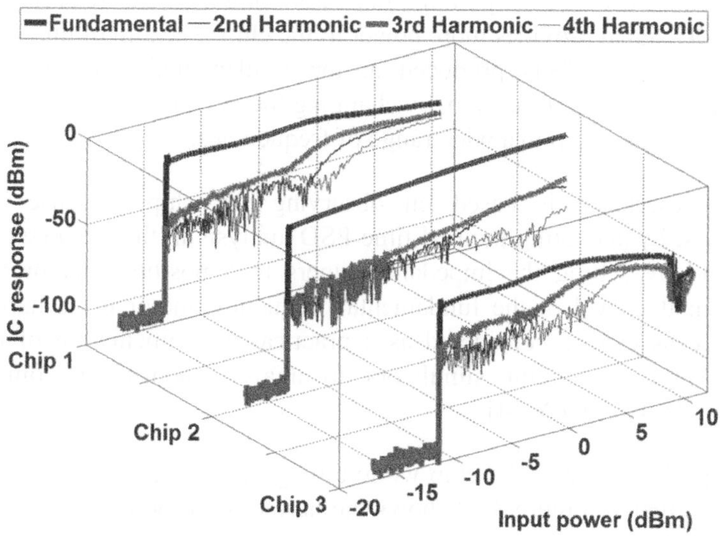

Figure 3.31. *Harmonic responses measured for the three chips for a sweep of power. A characterization until the fourth harmonic is presented*

Chip	Fund. (dBm)	Second harm. (dBm)	Third harm. (dBm)	Fourth harm. (dBm)
Chip 1	–21.9	–66.7	–61.3	–76.2
Chip 2	–19.9	–66.5	–61.6	–79.4
Chip 3	–25.4	–60.8	–49.3	–68.5

Table 3.5. *Harmonic responses from RFID chips*

3.4.4.1. *Harmonic treatment tests*

In this section, a treatment of the non-linearities is proposed by using an impedance tuner known as an antenna impedance emulator, as shown in Figure 3.24, and hereafter called the *antenna part*. In the design process of RFID tags, only the impedance value at the fundamental frequency is considered [AND 13]. Theoretically, this is enough to ensure the operation at the desired frequency. However, no special attention is paid to the harmonics produced by the chip. RF criteria state that the harmonics should be filtered, for example, in order to increase the purity of the RF-to-DC conversion performed by the rectifier part and thus providing a greater read range since a clean DC signal provides more power to the chip. The aim of this section is to evaluate the effects produced at the fundamental frequency by the non-linear behavior of the chip, by changing the matching condition between *antenna part* and chip in a wide range of frequencies.

The experience is based on observing the chip response and its harmonics, by activating the real-time PSD analysis in the DSO (see Figure 3.28) and by setting the impedance tuners to a position that produces a change in the level of the fundamental frequency and/or in its harmonics. This process can be considered as a practical experimentation of the HB method used in circuit simulators to analyze non-linear components [AGI 06, AGI 11, KEY 14].

The performance of the chip while tuning the impedance of the *antenna part* is evaluated. Figure 3.32 shows the harmonic responses of chip 1 until the fourth harmonic, in a scenario where the harmonics are increased by tuning the impedance of the *antenna part*. This is achieved at the expense of a 1.5 dB reduction on the chip sensitivity and 7 dB reduction on the chip response at the fundamental frequency. Table 3.6 compares the results before and after the harmonic treatment.

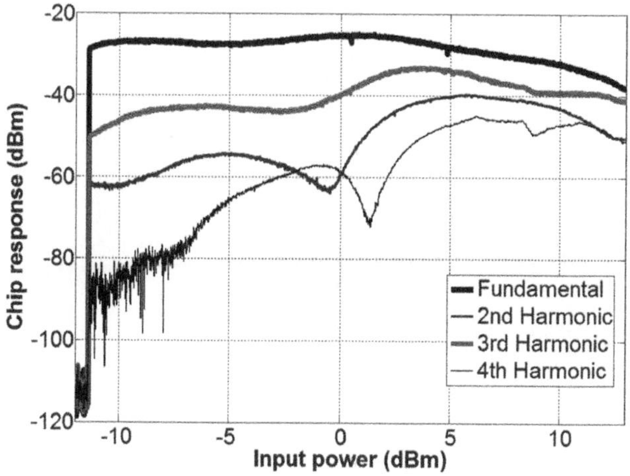

Figure 3.32. *Harmonic characterization for chip 1 after the harmonic treatment*

Chip 1	Sensit. (dBm)	Fund. (dBm)	Second h. (dBm)	Third h. (dBm)	Fourth h. (dBm)
No harmonic treatment	–12.9	–21.9	–66.7	–61.3	–76.2
Harmonic treatment	–11.4	–28.9	–61.3	–50.2	–87.6

Table 3.6. *Harmonic treatment*

Figures 4.11 and 4.12 show the input impedance of chip 1 in both modulation states after the harmonic treatment for the fundamental and its third harmonic, respectively. These measurements were performed at the new activation power. At the third harmonic, the resistive part decreases, and the reactance becomes more capacitive compared to the impedance at the fundamental frequency. Moreover, the difference between the modulation states is lower which means that, in a radiating scenario, the ΔRCS [SKA 09] will be smaller than the one at the fundamental frequency.

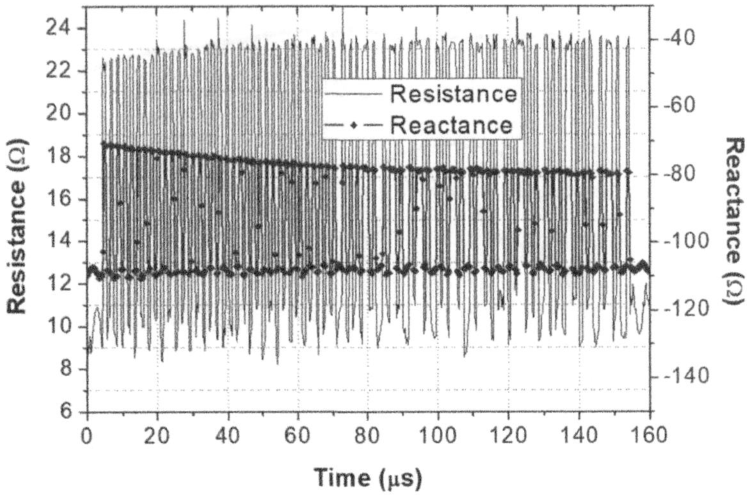

Figure 3.33. *Measured chip input impedance at the fundamental frequency in a temporal sweep. Scavenging and reflecting states can be seen*

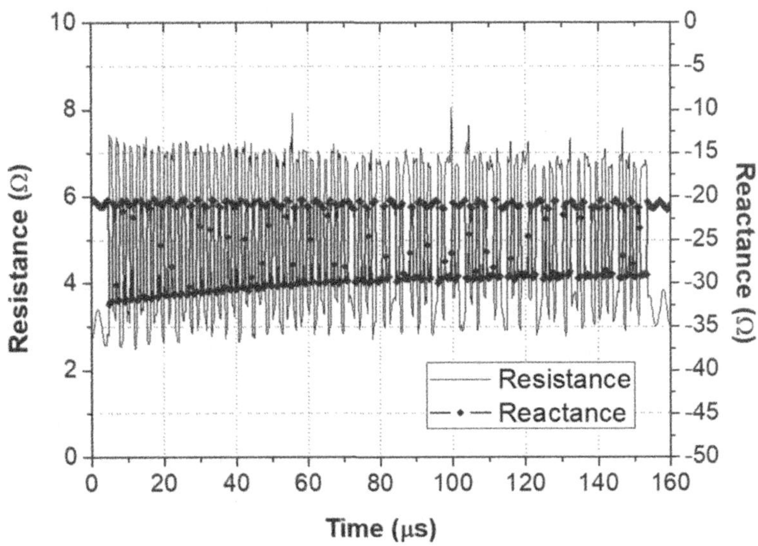

Figure 3.34. *Measured chip input impedance at the third harmonic frequency in a temporal sweep. Both states of modulation can be distinguished*

The results presented in Table 3.6 clearly show that an increase in the second and third harmonic results in slightly diminished performance at the fundamental frequency.

3.4.5. *Result exploitation*

After the chip is completely characterized, it is possible to export the scattering parameters measured with the RFID-NTP for a wide frequency range. Therefore, antenna designers can use the scattering parameters in order to optimize the antenna design using an electromagnetic CAD tool and ensure, for example, the complete filtering of backscattered harmonics. Indeed, the complex reflection coefficient of the RFID tag $\tilde{\rho}$, defined in [3.9], can be calculated for each impedance state [NIK 07, BOL 10]:

$$\tilde{\rho} = \frac{Z_c - Z_A{}^*}{Z_c - Z_a} \qquad [3.9]$$

where $Z_a = R_a + jX_a$ is the complex antenna impedance and Z_C is the complex RFID chip impedance in one state (scavenging or reflecting).

Therefore, a trade-off or independent optimization of the read range (load factor (LF) defined in [3.10]) and/or the ΔRCS (modulation efficiency (ME) defined in [3.11]) of the RFID tag can be performed at the design stage [BOL 10]:

$$LF = 1 - \left|\tilde{\rho}_{sca}\right|^2 \qquad [3.10]$$

$$ME = \left|\tilde{\rho}_{ref} - \tilde{\rho}_{sca}\right|^2 \qquad [3.11]$$

where $\tilde{\rho}_{sca}$ and $\tilde{\rho}_{ref}$ are the complex reflection coefficients of the tag in the scavenging and reflecting states, respectively.

On the contrary, it is interesting to note that both states of modulation are visible even at the third harmonic. This observation validates the hypothesis of redundant information originating on the same chip. Therefore, the design of a tag antenna radiating the fundamental signal and the third harmonic can be envisaged using the same LF and ME design parameter approach. It is worth noting the two main drawbacks to consider in the harmonic exploitation: (1) the low power of the third harmonic response and (2) the small ΔRCS.

3.5. Summary

The uncontrolled non-linearity in the current UHF RFID technology allows the tag to generate harmonic signals carrying information without constraints to the EPC Global Class-1 Gen-2 standard. The analysis made in this chapter considers the entire tag as the EUT (antenna and RFID chip), i.e. radiated measurements, and also the chip alone, i.e. conducted measurements.

Methods to configure the RFID air interface and to ease the spectral analysis of the tag response were described and later on used for the radiated and conducted chip characterization. The measurement procedures presented here introduce vector information about the signal allowing an evaluation of the RFID chip impedance in full operation mode and in a large frequency range.

The performance of the RFID chip and the real effect of the harmonics were evaluated by measuring harmonic compensation with a practical emulation of the antenna impedance. The harmonic treatment experiments would be the analogy of the HB method used for the simulation of non-linear circuits.

Subsequent lines of research can be undertaken by properly filtering the harmonics to improve the read range and/or the differential radar cross-section with the benefit of generating the whole RFID tag in existing applications. The analysis of the non-linearity of UHF chips presented here can be extended to any kind of Class-1 Generation-2 RFID chip, and it can be taken into account for the tag antenna design.

Besides, and contrary to traditional approaches, additional efforts to exploit the third harmonic response by the specialized tag antenna design promise new applications of and improvements to the existing ones, allowing a redundant communication from the tag to the reader. Nevertheless, the ability of the reader to collect and process harmonics should be enhanced to make this exploitation a reality. These topics together with experimental and parametric analysis are the core of Chapter 5.

4

Modeling the Harmonic Signals Produced by RFID Chips

The main objective of this chapter is to perform an in-depth study of the non-linear behavior of passive RFID chips, and to show that the produced non-linearities can be exploited to benefit the operational techniques of passive UHF RFID systems. A theoretical analysis of the harmonic production from passive RFID chips and the effect of the matching network and the tag antenna on the backscattering harmonic is performed. A new harmonic tag model that offers a maximum profit from the signals generated by the chip at fundamental and third harmonic frequencies is proposed. The theoretical analysis is confronted with radiated and conducted measurements.

4.1. Introduction

As long as the antenna and the chip are independently designed [RAO 50], and due to the non-linear behavior of the RFID chip [NIK 09, DEV 05, BOL 10], the tag antenna will radiate some of the harmonic currents of the CW that feed the chip. These harmonics are also carrying the impedance modulation effect, notably producing then a redundant backscattering modulation. The underlying theoretical basis of these phenomena was explained and experimentally demonstrated in Chapter 3 by means of radiated measurements considering the whole tag and by conducted measurements in which a test bench was directly connected to the RFID chip.

In order to complete the experimental approach presented in Chapter 3, this chapter aims to perform a theoretical analysis of harmonic signals produced by the RF section of passive RFID tags. The analysis pinpoints in the operation of a traditional tag working at the fundamental frequency of the system (f_0), and also in the operation of a new tag able to enhance the backscattered signal strength at its third harmonic ($3f_0$), hereinafter the so-called harmonic tag (HT).

The chapter explains the harmonic production of passive RFID chips, and the effects of the matching network and tag antenna on harmonic backscattering phenomena. The theoretical analysis is performed in three stages: (i) the calculation of a general expression for the current in the rectifier part of RFID chips; the expression comprises fundamental and harmonic components; (ii) the study of the matching network used in traditional RFID tags in order to evaluate the harmonic backscattering strength; (iii) the study of a new matching network architecture that enhances the backscattering strength at $3f_0$ maintaining the operation at f_0, therefore constituting the HT operation theory. The analysis presented in this section uses lumped components to model the operation of a distributed impedance matching network integrated in antennas used in passive RFID.

4.2. Analysis of harmonic currents in RFID chips

4.2.1. *Review of Dickson analysis*

The most extended architecture used to study the rectification process in UHF RFID is the Dickson charge pump and its derived equations [DIC 76, USA 04, BAR 06, CUR 05]. Although the rectifier architecture used in this chapter is very similar to the Dickson charge pump, it has some important differences [BAR 09]. Since UHF tags operate with significantly lower signal levels and higher frequency than the original Dickson analysis, some modifications must be made to the RF-to-DC equation proposed by Dickson. The original Dickson analysis assumes constant forward voltage drop of the rectification device, the so-called "threshold voltage", although this forward voltage drop is actually a function of the load current of the rectifier. This variation is not significant for large voltage swings across the diode as is the case for the charge pump application. However, in passive UHF RFID, the

RF input voltage is very small, on the same order of the forward voltage drop. As a result, the non-linear forward voltage drop has significant impact on the RF-to-DC equation. A parametric non-linear analysis theory has been developed to incorporate the effect of the non-linear forward voltage drop. It was first applied to predict the RF-to-DC conversion for single-diode rectifiers in [HAR 92] and was later expanded to multi-stage rectifiers in [DEV 05].

The multi-stage voltage multiplier model and the analysis method used in this chapter are supported by several works that use similar architectures to predict the RF-to-DC conversion for rectifiers in UHF RFID applications, being aware of the low RF input voltage present in this technology [DEV 05, KAR 03, DEV 05, SAN]. Figure 4.1 shows the RF section of a passive RFID tag with a two-stage voltage multiplier, hereinafter referred to as rectifier. The rectifier circuitry is composed of a cascade of N-stage voltage multipliers.

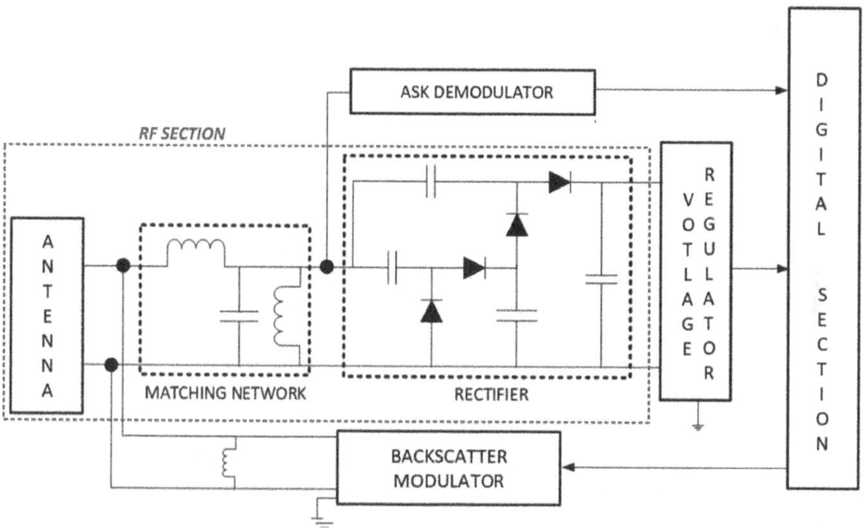

Figure 4.1. *Passive tag architecture*

Because of the non-linear effects of the rectifier part of the RFID chip, the current in the antenna I_{ant} presents components at frequencies that are integer multiples of the operation frequency. Indeed, in the scavenging state, the input voltage induced in the tag antenna and then conducted to the rectifier at f_0, generates an input current I_{in} dominantly constituted by the odd harmonic components of f_0. Figure 4.2 shows the Schottky diode model and its different configurations for the analysis.

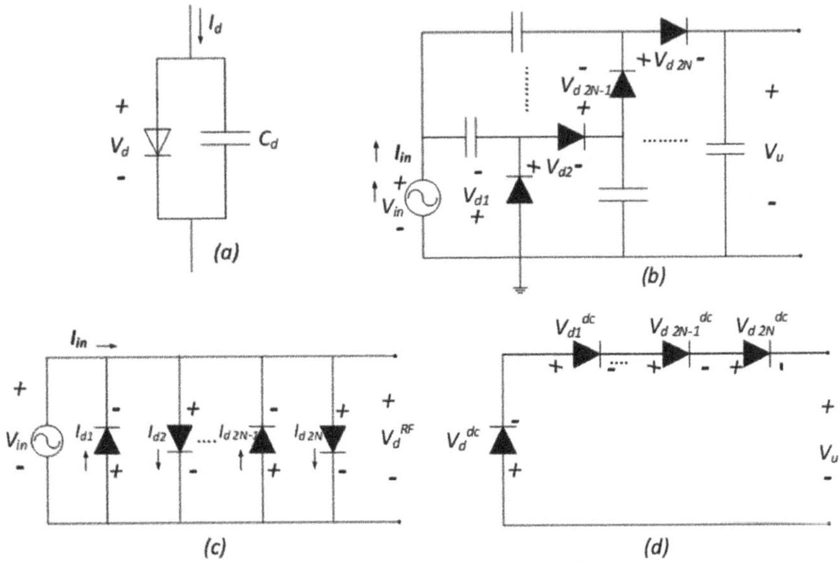

Figure 4.2. *Equivalent circuit of the Schottky diode for the analysis in DC and RF. a) Equivalent circuit of the Schottky diode. b) Equivalent rectifier circuit under the steady state. c) RF equivalent circuit: the capacitors become short circuits and the 2N diodes are in parallel. d) DC equivalent circuit: the capacitors do not allow the current to pass through and the diodes are in series*

4.2.2. Calculation of the harmonic currents

In order to determine the general expression of the input current to the rectifier, I_{in} in Figure 4.2(b), it is necessary to introduce the current–voltage expression of the diode:

$$I_d = I_s(e^{\frac{V_d}{V_T}} - 1) + C_d \frac{dV_d}{dV_t} \quad [4.1]$$

where I_s is the saturation current of the diode, V_d is the voltage at the diode terminals and V_T is the thermal voltage, which is approximately 25.8 mV at 300 K. $V_T = kT/e$, where k is the Boltzmann constant, e is the electron charge and T is the temperature. C_d is the junction capacitance of the diode. It should be noted that this model does not consider the parasitic effects of the package.

If the capacitors in the rectifier are well-dimensioned, with V_{in} being a sinusoidal voltage with a frequency f_0 and an amplitude V_0 at the input of the rectifier, it is reasonable to consider that the rectifier presents the following equivalent circuits for RF signals and DC signals:

– RF analysis: the capacitors are short circuited, the RF voltage at each diode is $V_d^{RF} = \pm V_0 \cos(\omega_0 t)$ and all the diodes are in parallel (ω_0 stands for the angular frequency $\omega_0 = 2\pi f_0$). The sign of V_d^{RF} depends on the orientation of the diode.

– DC analysis: the capacitors are open circuited and all the diodes are in series. The DC voltage is equally divided in each diode $V_d^{dc} = -\frac{V_u}{2N}$, where V_u is the output voltage.

Then, the voltage V_d in each diode is:

$$V_d = V_d^{RF} + V_d^{dc} = \pm V_0 \cos(\omega_0 t) - \frac{V_u}{2N} \qquad [4.2]$$

and substituting [4.2] in [4.1] yields

$$I_d = I_s(e^{\frac{V_0 \cos(\omega_0 t) - \frac{V_u}{2N}}{V_T}} - 1) \pm C_d \frac{d(V_0 \cos(\omega_0 t))}{dV_t} \qquad [4.3]$$

In order to calculate I_{in} for an N-stage rectifier, it is necessary to calculate the current in a single stage node and then multiply it by N, yielding

$$I_{in} = NI_s\left[e^{-\frac{V_u}{2NV_T}}\left(e^{\frac{V_0 \cos(\omega_0 t)}{V_T}} - e^{-\frac{V_0 \cos(\omega_0 t)}{V_T}}\right)\right] - 2NC_d V_0\, sin(\omega_0 t) \qquad [4.4]$$

The exponential of a cosine function in [4.4] can be expanded using the Fourier series

$$e^{\pm X\cos(\omega t)} = B_0(\pm X) + 2\sum_{n=1}^{\infty} B_n(\pm X)\cos(n\omega t) \qquad [4.5]$$

where $B_n(X)$ is the modified Bessel function of the first kind, of order n and argument X. The input current generated by a sinusoidal input voltage at the operation frequency can be written using [4.5] in [4.4], yielding [4.6], which is the current–voltage expression of a 2N-stage rectifier circuit:

$$I_{in} = 4NI_s e^{-\frac{V_u}{2NV_T}} \sum_{n=1}^{\infty} B_{2n-1}\left(\frac{V_0}{V_T}\right)\cos[(2n-1)\omega_0 t] - 2NC_d V_0 \sin(\omega_0 t) \qquad [4.6]$$

From [4.6], it is observed how the even RF components of ω_0 are eliminated due to the opposite sense of currents in each diode at each voltage doubler stage (RF analysis shown in Figure 4.2 (c)). In general, this phenomenon explains the odd harmonic predominance in voltage doublers. In addition, in an N-stage rectifier, the total capacitance, hereinafter called C_{eq}, is equal to $2NC_d$, and if the matching network (Figure 4.1) is well-dimensioned, the current contribution through internal capacitors C_d in [4.6] is eliminated since C_{eq} is compensated by the shunt inductance of the matching network. The operation and design of the matching network are detailed in the next section.

The first components of the input current of the rectifier described in [4.6] are shown in Figure 4.3 in the logarithmic scale when using representative values $N = 6$, $V_T = 0.025\,V$, $V_u = 1\,V$, and $I_S = 10^{-7}$ A [AAZ 09]. It can be seen that the amplitude current component at f_0 (i.e. I_{10}) and the amplitude current component at $3f_0$ (i.e. I_{30}) are close to each other.

Focusing the analysis on the components at f_0 and $3f_0$, Figure 4.4 shows the ratio I_{30}/I_{10} in the logarithmic scale. The ratio is an intrinsic consequence of the non-linear nature of diode-based devices. From [4.6], the rectifier produces harmonic currents which the ratio does not depend on the parameters N, V_T, V_u, I_S but only on V_0. From Figure 4.4, it can be seen, for instance, when $V_0/V_t \geq 6$ ($V_0 \geq 0.15\,V$), the amplitude of I_{30} is half of the amplitude of I_{10} and they become closer as V_0 increases. For values of $V_0 \geq 0.44\,V$, the difference between I_{30} and I_{10} is less than 1 dB. The non-linear effect in rectifier devices is higher when V_0 increases, which leads to a current amplitude at $3f_0$ closer to the current amplitude at f_0.

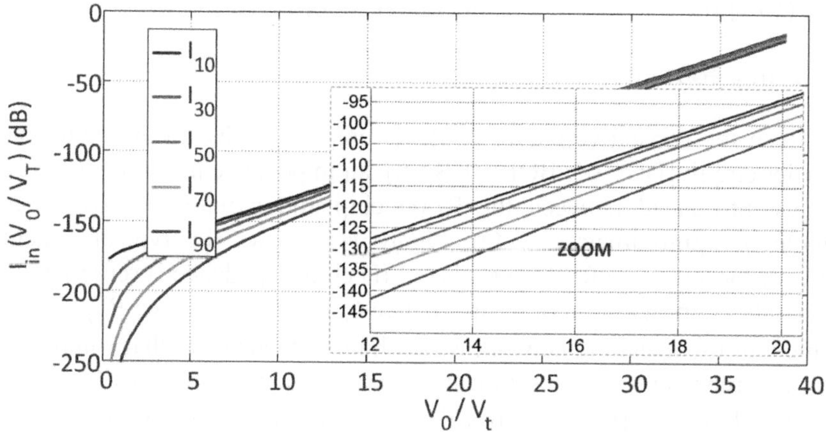

Figure 4.3. *Current–voltage curve for a rectifier of N voltage-multiplier stages, modeled using the modified series of Bessel. The total current is composed only of the odd harmonic components. A similarity between the current amplitude at f_0 and current amplitude at $3f_0$ can be noted. For a color version of this figure, see www.iste.co.uk/andia/non-linearites.zip*

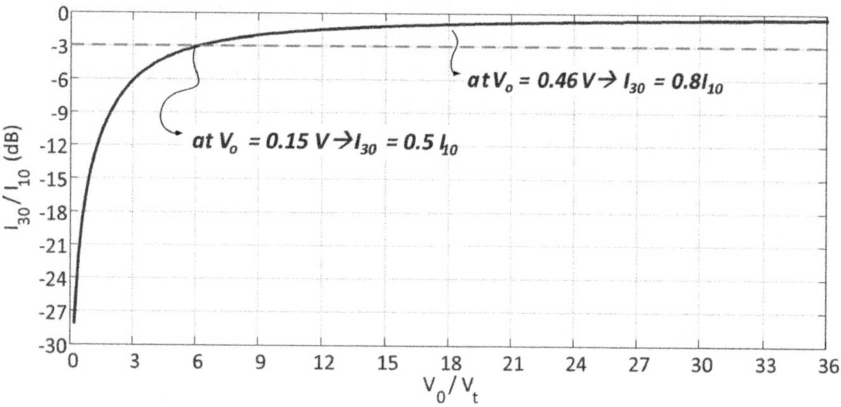

Figure 4.4. *Comparison of I_{30}, the amplitude of the current in the rectifier at $3f_0$, and I_{10}, the amplitude of the current in the rectifier at f_0. At $V_0/V_t = 6$ ($V_0 = 0.15$ V), I_{30} is the half of I_{10} which becomes closer as V_0 increases*

4.3. Third harmonic in traditional RFID tags

4.3.1. *Impedance matching network for f_0*

Given the capacitive nature of RFID chips at f_0 due to the diode-based rectifier circuit [DEV 05, BOL 10, SAN], an impedance matching network is always introduced between the RFID chip and a pure resistive antenna (dipole of approximately 73 Ω). Figure 4.5 shows the typical matching network for a passive RFID tag. The RFID chip impedance is represented by the parallel configuration of a capacitance C_{eq} and a resistance R_{eq}. The first element of the matching network is the shunt inductance L' that compensates C_{eq} and represents the traditional inductive loop used in RFID antenna design [MAR 08] (C_{eq} describes the capacitive nature of the RFID chip). Once the chip reactance is canceled, the remaining high shunt resistance R_{eq} (approximately 1.8 kΩ [IMP]) needs to be transformed to the comparatively low series resistance of a dipole antenna, hereinafter R_A [BEH 98]. The LC network is often used to perform the transformation as follows: (i) C transforms the shunt resistance of the RFID chip R_{eq} to a smaller series value matching it to R_A, being $C = Q/(\omega_0 R_{eq})$; (ii) L cancels C, with $L = QR_A/\omega_0$. The parameter Q defines the quality factor of the LC network and it is bounded to the resistance transformation ratio between chip and antenna. In other words, Q defines the performance of the LC network as a function of the power transferred between the antenna and the chip, and vice versa at f_0, which is given by

$$Q = \sqrt{\frac{R_{eq}}{R_A} - 1} \qquad [4.7]$$

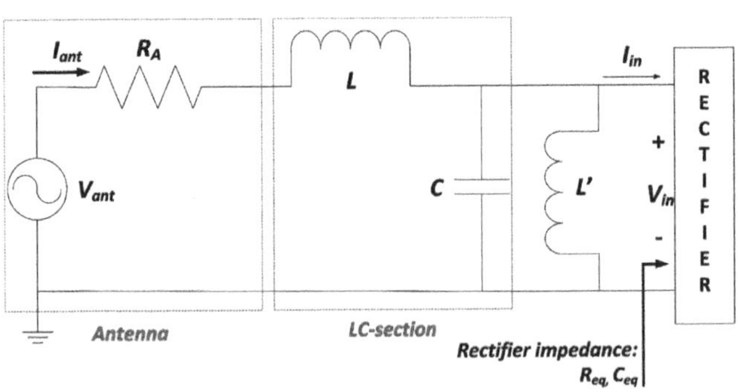

Figure 4.5. *Equivalent circuit of the RF section of a passive RFID tag*

Figure 4.6 shows the transmission coefficient of the LC network for different values of Q, keeping R_{eq} fixed. A higher Q ensures a better power matching at 868 MHz, therefore higher isolation at frequencies different from f_0.

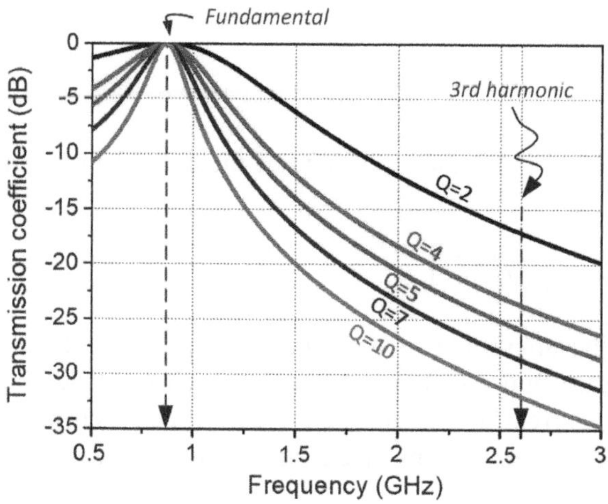

Figure 4.6. *Transmission coefficient of the LC network with different Q values. For a color version of this figure, see www.iste.co.uk/andia/non-linearites.zip*

It is worth noting that the assumption of capacitive impedance for the chip (R_{eq} and C_{eq}) and pure resistive impedance for the antenna (R_A) is realistic. Indeed, passive RFID chips and rectifier diode-based circuits have in general capacitive impedance, typically of the order of $20 - j150$, approximately 868 MHz [MAR 08, IMP, NXP]. Regarding the antenna, the analysis considers a resonant dipole with a dimension close to a half-wave length (i.e. zero imaginary component at the resonant frequency).

4.3.2. *Influence of Q in the backscattered signal at 3f₀*

The effect of Q on the signal radiated by traditional tag antennas at $3f_0$ is modeled using the architecture shown in Figure 4.5. With V_{ant} being the open circuit voltage of the tag antenna and V_{in} the input voltage of the rectifier, both at f_0, and assuming a lossless matching network, the delivered input power, equal to V_{ant}^2/R_A, must be equal to its output power, V_{in}^2/R_{eq}. The input power of the matching network is $P_{ant} = V_{ant}\, I_{ant}$, and the output power of the matching network is $P_{eq} = V_{in}\, I_{in}$, where I_{ant} is the current at the tag antenna at f_0, and I_{in} is defined in [4.6]. Therefore, it is also valid to say

$$\frac{I_{in}}{I_{ant}} = \sqrt{\frac{R_A}{R_{eq}}} \qquad [4.8]$$

From [4.7] and [4.8], the amplitude of the current in the antenna I_{1ant} at f_0 is given by

$$|I_{1ant}| = I_{10}\sqrt{Q^2+1} \qquad [4.9]$$

where I_{10} is the amplitude of the rectifier input current I_{in} at f_0 given by [4.6]. In order to find a similar relation between I_{3in}, the component of the rectifier input current at $3f_0$, and I_{3ant}, the reflected current in the antenna terminals at $3f_0$, an equivalent model of the tag backscattering the harmonic is presented in Figure 4.7.

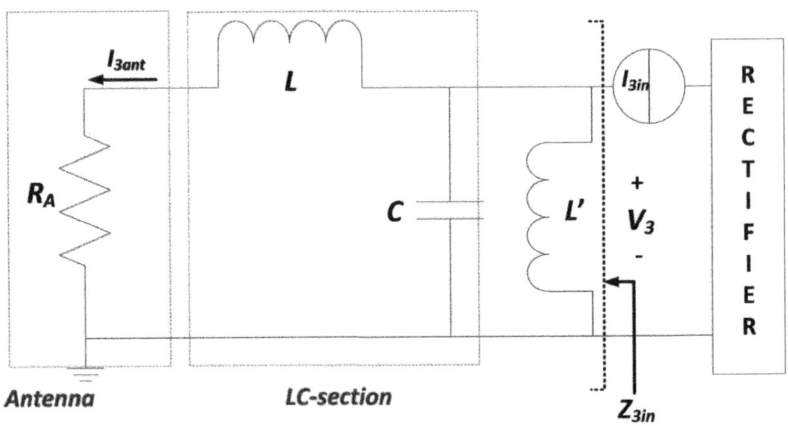

Figure 4.7. *Equivalent circuit of the tag when the current at $3f_0$ is being reflected in the antenna*

Let V_3 be the shunt voltage when the current I_{3in} at $3f_0$ is generated by the rectifier, as shown in Figure 4.7; therefore, $V_3 = I_{3in}\, Z_{3in}$. Z_{3in}, the input impedance seen by the rectifier, is given by

$$Z_{3in} = \frac{-\omega^2 LL' + j\omega^2 L' R_A}{R_A(1-\omega^2 L'C) + j\omega(L' - \omega^2 LL'C + L)} \qquad [4.10]$$

V_3 must be equal to the voltage $I_{3ant}(R_A + j\omega L)$, which yields $I_{3in}Z_{3in} = I_{3ant}(R_A + j\omega L)$. Substituting the relations between C, L, R_A, R_{eq} and Q considered above and $\omega = 3\omega_0$, it is possible to calculate the amplitude of the current in the antenna I_{3ant} at $3f_0$ due to the current I_{3in} generated by the rectifier at $3f_0$, as

$$|I_{3ant}| = \frac{3I_{30}L'\omega_0(Q^2+1)}{\sqrt{(R_{eq} - 9\omega_0 L' Q)^2 + 9(QR_{eq} + (1-8Q^2)\omega_0 L')^2}} \qquad [4.11]$$

where I_{30} is the amplitude of I_{3in}, the component of the rectifier input current I_{in} at $3f_0$ in [4.6]. From the capacitive nature of the RFID chip, the resistance conversion of the matching network from a high shunt resistance R_{eq} (approximately 1.8 $k\Omega$), to a comparatively lower series resistance R_A (approximately 73 Ω), imposes that $C \gg C_{eq}$. Since L' resonates with C_{eq} at the f_0, then $L' = \frac{1}{C_{eq}\omega_0^2}$. The following statement consequent with [DEV 05] is then derived:

$$Q\omega_0 L' \gg R_{eq} \qquad [4.12]$$

Using [4.12], it is possible to rewrite [5.11] as follows:

$$|I_{3ant}| = \frac{I_{30}(Q^2+1)}{\sqrt{64Q^2 - 7Q^2 + 1}} \qquad [4.13]$$

The power delivered to the tag antenna terminals at f_0 and at harmonic frequencies is differently affected by the matching network and can be expressed in terms of Q. From [4.9] and [4.13], it is possible to calculate the power delivered to the tag antenna terminals at f_0 and $3f_0$, P_1 and P_3 respectively as follows:

$$P_1 = \frac{1}{2}|I_{1ant}|^2 R_A = \frac{I_{10}^2(Q^2+1)R_A}{2} \qquad [4.14]$$

$$P_3 = \frac{1}{2}|I_{3ant}|^2 R_A = \frac{I_{30}^2(Q^2+1)^2 R_A}{2(64Q^2-7Q^2+1)} \qquad [4.15]$$

Figure 4.8 shows the law decay of the power P_3 at $3f_0$ with respect to the power P_1 at f_0, as a function of Q and the ratio V_0/V_t. Two major patterns can be distinguished:

– the power level at $3f_0$ approaches the power level at f_0 as the input voltage to the rectifier V_0 increases;

– the power level at $3f_0$ approaches the power level at f_0 as Q decreases.

It is expected that with a higher Q, the rectifier input voltage at f_0 is also higher, which leads to a greater non-linear effect, thereby causing the rectifier to produce a higher component at $3f_0$. However, a higher Q also ensures a higher attenuation at frequencies different from f_0. To summarize, when Q is higher, so, is the suppression of the power transmitted from the chip to its antenna at $3f_0$.

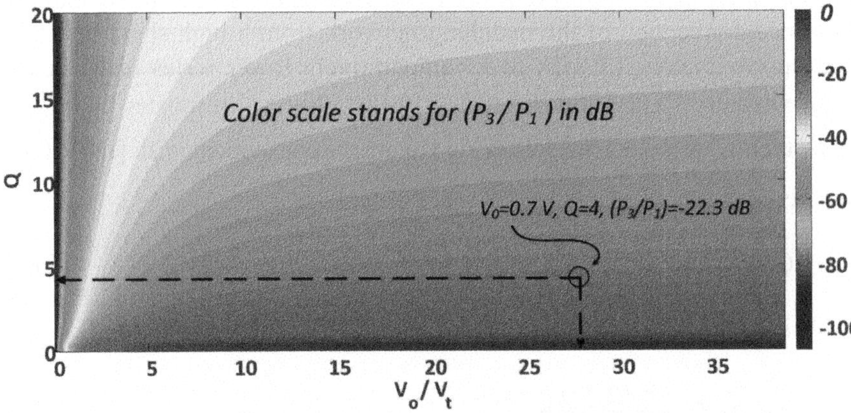

Figure 4.8. *Law decay of the power P_3 at $3f_0$ with respect to the power P_1 at f_0 as a function of Q and V_0/V_t for traditional tags. For a color version of this figure, see www.iste.co.uk/andia/non-linearites.zip*

The harmonic power can be potentially radiated if the antenna structure allows it, i.e. profiting from the current distribution of harmonics in a specific antenna structure [AND 13]. According to the proposed model, a traditional RFID tag having a dipole antenna structure, which has a current distribution at $3f_0$ different from zero at the input and a distributed matching network with $Q = 4$, radiates a signal at $3f_0$, i.e. 22.3 dB, which is lower than the signal at f_0. This model explains why certain commercial tags backscatter a signal at $3f_0$ with stronger level than others, certainly depending on the tag antenna structure as evidenced in section 3.3 of Chapter 3 for the tag Rafsec MemoryStick [UPM].

4.4. How to profit from the third harmonic signal

4.4.1. *Dual-band impedance matching network*

The analysis presented in this section develops the operation theory of the HT that extends the equivalent model presented in Figure 4.7. A new matching network that could enhance the backscattering strength of HTs at $3f_0$ is presented in Figure 4.9. The new matching network composed of the cascading of two LC networks [ALI 07] is characterized by a new quality factor Q^* calculated as

$$Q^* = \sqrt{\left(\frac{R_{eq}}{R_A}\right)^{1/2} - 1} \qquad [4.16]$$

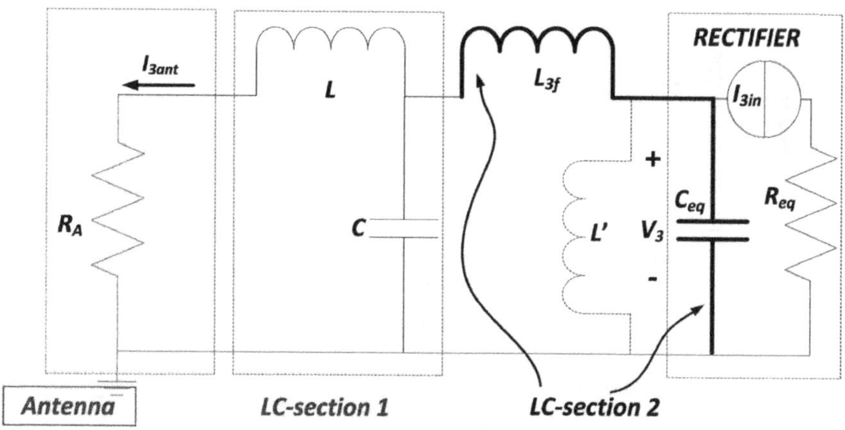

Figure 4.9. *Proposed matching network to exploit the signal generated by the RFID chip at 3f₀. The cascade LC matching network allows the chip to transmit a maximum power towards the antenna at 3f₀. L' is always considered in the analysis*

The new element introduced is the series inductance L_{3f} that resonates with both the original LC network and C_{eq} at $3f_0$. The new matching network should be dimensioned choosing an L_{3f} which does not affect the traditional operation at f_0, i.e. L_{3f} should act as a short circuit at f_0. In the same way, the dimensioning criteria have to consider that the shunt inductance L' should not intervene in the operation at $3f_0$, i.e. L' should act as an open circuit at $3f_0$. The first step to choose the L_{3f} value is to calculate the inductance necessary to compensate the chip capacitance at $3f_0$ (inductive matching), and depending on the values of the original LC network, L' should be tuned to resonate at $3f_0$. Realistic values for the proposed model are $C_{eq} = 1.2\,pF, R_{eq} = 1.8\,K\Omega$, $L' = 27.8\,nH$, $L_{3f} = 10.6\,nH$, $C = 0.6\,pF$, $L = 52.8\,nH$ and $R_A = 73\,\Omega$.

Figure 4.10 shows the transmission coefficient of the proposed matching network for the HT, compared to the matching network of a traditional tag.

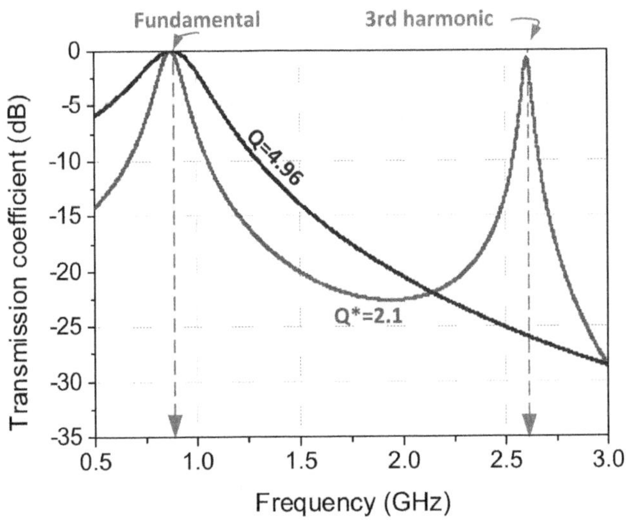

Figure 4.10. *Transmission coefficient of the cascade LC network of the HT antenna (Q* = 2.01) compared to an original LC network in traditional tags (Q = 4.96)*

4.4.2. Backscattered signal at $3f_0$ by the HT

From [4.8] and [4.16], the amplitude of the current at $3f_0$ at the HT input I_{3ant} can be written as

$$I_{3ant} = I_{30}(Q^{*2} + 1) \quad [4.17]$$

and from [4.7] and [4.16], the relation between Q^* and Q is

$$Q^* = \sqrt{\sqrt{Q^2 + 1} - 1} \quad [4.18]$$

Substituting [4.18] in [4.17], the current in the antenna at $3f_0$ is

$$|I_{3ant}| = I_{30}\sqrt{Q^2 + 1} \quad [4.19]$$

From [4.9] and [4.19], it can be observed how the new matching network treats the current of the chip at f_0 and $3f_0$ in a similar manner.

Using [4.9] and [4.19] in [4.14] and [4.15], respectively, the power radiated by the HT at f_0 and $3f_0$ harmonic can be expressed as:

$$P_{1HT} = \frac{I_{10}^2(Q^2+1)R_A}{2} \quad [4.20]$$

$$P_{3HT} = \frac{I_{30}^2(Q^2+1)R_A}{2} \quad [4.21]$$

Figure 4.11(a) shows the law decay of the power P_{3HT} at $3f_0$ with respect to the power P_{1HT} at f_0, as a function of Q^*, Q and the ratio V_0/V_t. If the matching network is well-designed, which means an antenna having a distributed matching network that cancels the RFID chip reactance at f_0 and $3f_0$, the difference between the backscattered power by the HT at $3f_0$ and f_0 depends only on the intrinsic non-linear behavior of the chip (Figure 4.4) and is shown in Figure 4.11(b) at a certain level of the input voltage V_0.

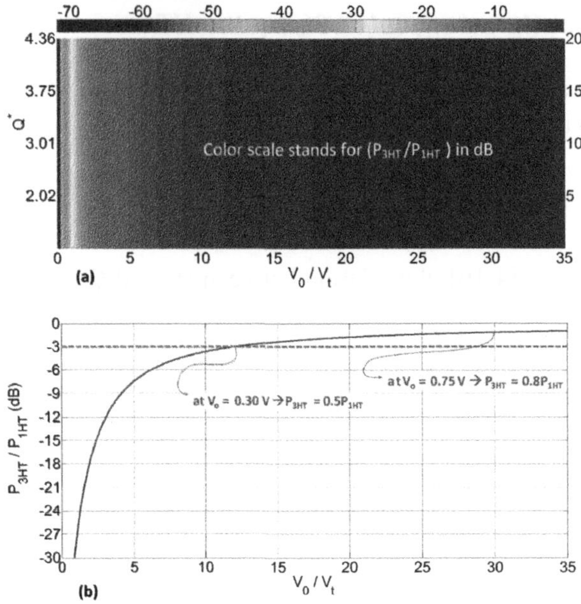

Figure 4.11. *Law decay of the power at $3f_0$ with respect to the power at f_0 for the HT. a) Law decay as a function of Q*, Q and V_0 for the HT. b) Law decay as a function of V_0. If the matching network is well dimensioned, the ratio depends only on the non-linearity of the chip. For a color version of this figure, see www.iste.co.uk/andia/non-linearites.zip*

Experimental tests reported in section 3.4 of Chapter 3 have shown the feasible enhancement of the signal generated by the chip at $3f_0$ using an antenna impedance emulator. The so-called harmonic treatments are based on emulating proper matching networks that could enhance the response at $3f_0$ compared to the case of an optimum emulated antenna at f_0 having the same chip. An enhancement of 11 dB on the response at $3f_0$ is reported to perform conducted measurements.

4.5. Summary

A theoretical analysis of the harmonics produced by the non-linear behavior of passive UHF RFID chips has been presented in this chapter. The analysis based on a modified version of the Dickson rectifier considers different input voltages and it is valid to analyze the chip behavior when low input signals such as the ones used in passive RFID are present.

This chapter has explained the particular predominance of even or odd harmonics depending on the architecture of the rectifier circuit, which determines the canceling or not of currents in the electrical analysis. Indeed, a single diode will produce all the harmonics with predominance of those ones of low order, i.e. the second harmonic. However, if the rectifier is composed of voltage doubler stages (where the number of diodes is a multiple of two), an odd harmonic predominance is present.

The antenna–chip impedance matching is treated with a lumped network that models the operation of antenna structures that includes a distributed matching network. The theoretical analysis is confronted with the operation of traditional tags reported in Chapter 3. A new harmonic tag antenna model that comprises a matching network that fully exploits the fundamental and third harmonic signals generated by EPC Class-1 Gen2 RFID chips is introduced. The analysis of this new HT is also confronted with experimental harmonic treatment tests reported in section 3.4 of Chapter 3 and confirms the feasible exploitation of the third harmonic signal of passive RFID chips. In order to give an example, the most intuitive application could be the exploitation of harmonic signals as information carriers that establish frequency diversity from tag to reader, without breaching the standard regulations. Such a concept establishes a potential foundation for new applications that the chip self-generated harmonic signal with redundant information can offer. This idea will be highlighted in Chapter 5 with several prototypes and also other possible concepts where the third harmonic is seen as an energy source.

5

Applications: Augmented RFID Tags

This chapter unfolds, from an experimental approach, the new concept of exploiting the harmonic signals generated by the non-linear behavior of traditional passive UHF RFID chips. In this chapter, two innovative ideas are explored considering the third harmonic as either an energy source or an information source carrying the identification signal. Specifications, methodology and experiments are explained for two kinds of augmented tags: (i) tags with harmonic communication using the EPC Class-1 Gen2 RFID protocol and (ii) tags with harmonic harvesting to feed supplementary low consumption electronics. Practical design aspects and evaluation metrics in compliance with standard regulations for EPC Class-1 Gen2 RFID are presented. The building of an equivalent low-profile circuit with RF performance similar to that of a commercial RFID chip is explained. The performance of the solutions is illustrated by simulation and measurements. In addition, a thorough parametric study is carried out for all the prototypes, especially with respect to the antenna structure, the kind of chip, the received power and the read range.

5.1. Introduction

As discussed in Chapter 1, RFID technology and UHF tags have reached a remarkable level of performance in terms of chip sensitivity, tag read range, inlay size and flexibility to fit a variety of items to be tracked. It is also a standardized technology that allows unitary identification and low cost of production. These very interesting features favor the implementation of RFID solutions in thousands of applications and domains such as traceability, logistics and access control. RFID has become ubiquitous in industry and our daily life (ticketing, payment, passports, car keys, etc.). It is also intensively used for health and hospital monitoring and personal assistance. However, in all these applications, the duty of the tag is to return an ID to the reader and achieve the unambiguous identification only. Over the last few years, with the development and generalization of wireless solutions, new concepts have entered the field of applications. Among these concepts, the paradigm of the Internet of Things (IoT) is probably the most effective in terms of application management and optimization of performance. The perspective of RFID in terms of new paradigms and the IoT is more and more under consideration worldwide. Indeed, passive UHF tags are based on RFID chips that require –22dBm only for activation. They exhibit read-ranges of more than 25 meters. These performances allow them to appear as an ideal candidate to implement the last few meters of the IoT. To do so, the tags must be augmented in terms of functionalities to perform more than identification. The transformation of RFID tags into augmented tags is very effective for wireless networks. Indeed, the use of the standardized RFID communication protocol will allow designers to exploit the normalized RFID commands to define new commands suitable for monitoring augmented tags. There are several ways to transform a UHF RFID tag into an augmented tag. They can be grouped into three categories. The first one will exploit the sensitivity of the tag antenna to the physical change that appears in its near-field region. Such a technique can be used for sensing purposes. The second category involves the integration of an external device to the RFID chip like a sensor. This second category is more related to the behavior of the RFID chip and the variation of its electrical response as a function of some external parameters such as RF power and temperature. In that case, we exploit the non-linear properties of the RFID chip. Different methods of tag augmentation, based on non-linearities of the RFID chip, are discussed in the following sections of this chapter. It is worth noting that the augmentation becomes interesting if the cost of tags is maintained low and the interrogation system is enriched and updated by additional commands to collect the parameters related to the new functionalities acquired by RFID tags.

5.2. Harmonic communication in passive UHF RFID

As partially introduced in Chapter 2, the study and exploitation of non-linearity in diode-based devices is not new. Approaches based on a one-bit system in which the harmonic or intermodular components serve to indicate tag presence without a specific communication protocol are known as harmonic radar. Used as a tracking solution, the invention was patented in 1975 [SHE 75] opening the path for several derived applications, e.g. in insect tracking [COL 04]. With the boom of RFID systems, the harmonic radar concept has been recently re-purposed. In [GOM 07, SON 14], a tag response is achieved by exploiting the intermodulation phenomenon of a diode. In [FAN 10], a specific non-linear RF circuit is developed with the aim of enhancing the harmonic strength compared with multiplier diodes used at 10 GHz. The circuit integrated to an antenna conforms to a harmonic transponder of one bit and later the transponder is used for accurate indoor ranging [YUN 14]. The authors of [RAS 14, RAS 14] have studied and generalized the design techniques for harmonic transponders based on a single detector or varactor diode. The concept was adopted for low-cost solutions using paper substrates and inkjet printing technology [ALI 10, ORE 11] or even an eco-friendly solution using an organic diode [VIR 14]. Another exploitation idea [ROM 09] proposes the detection of counterfeit RFID cards based on their electromagnetic characteristics rather than the digital information that they transmit. In this approach, one of the stages considers the electromagnetic signature that harmonic signals produce. Electromagnetic energy harvesting is an emerging field where the non-linearity of diode-based devices is also being exploited. The authors of [TRO 09, COL 14, BOA 13, BOR 14] have proposed power-optimized waveforms to improve the range and reliability of RFID systems. By using specially designed signal waveforms at the transmitting terminal that somehow excite the non-linear diode behavior of rectifying diodes in a more efficient manner, the RF-to-DC conversion efficiency of the diode can be increased. In the same field, the work reported in [LAD 14] proposes a millimeter-wave rectenna at 35 GHz that also harvests the harmonic signals generated by its rectifier. The proposed solution enhances the RF-to-DC conversion efficiency.

In the specific field of passive UHF RFID, the experimentation on backscattered harmonics was first introduced in 2009 [NIK 09], but no additional research efforts were followed, probably because of the evidenced low power level and limitations on the characterization method.

Contrary to the conventional paradigm in RF communication circuits where non-linearities are undesired and their level is highly reduced [DEV 05, BOL 10], this section develops the concept of a new communication function by exploiting non-linearities of passive EPC Class-1 Gen2 RFID chips. The exploitation of harmonic signals as information carriers establishes frequency diversity from tag to reader, without breaches in standard regulations. The benefit of the frequency diversity is the possibility to use a spectrum which has much reduced self- and multi-path interference, therefore countering multi-path and leakage effects at reception, but not a read range enhancement. The developed concept establishes a foundation for new potential applications that the self-generated harmonic chip signal with redundant information can offer. With such a motivation, this chapter aims to meet two specific objectives:

– to define exploitation policies when the harmonic communication between the harmonic tag (HT) and the reader is envisaged. The policies treat regulation definitions, considerations on the reader side, metrics to evaluate the HT performance and design guidelines for the HT;

– to explain the HT design and operation with a real HT design example and a deep parametric study at different powers, frequencies and with respect to read range performance.

The theoretical analysis presented in Chapter 4 has shown the possibility of increasing the backscattering strength at $3f_0$, i.e. to exploit the harmonic generation of passive RFID chips by properly matching the impedance of the antenna. This section aims to present considerations on the use of HTs in a tag-to-reader harmonic communication regard. Indeed, it is intended to declare the diversity in frequency where signals from tag-to-reader at f_0 and at $3f_0$ carry the same information. Aspects such as deployment regulations, harmonic reader operation considerations, HT evaluation metrics and HT design guidelines are unfolded.

5.2.1. *A review of the regulations*

5.2.1.1. *Regulations at f_0*

In order to make possible and favor the use of the HT, e.g. regarding the compatibility with other systems, the parent design policy is to respect current standards and local regulations on passive UHF RFID [EPC, ETS 08].

In that regard, since dual-band operation is desired (at f_0 from the passive UHF RFID system, i.e. in Europe 868 MHz and at $3f_0$, i.e. 2604 MHz), reader and tag spectrum mask regulations must be accomplished.

No change is envisaged on the transmitter side of the RFID reader. In other words, HTs should operate with RFID readers that comply with the spectrum mask regulations [EPC 15]. In the same manner, the tag spectrum mask defined in [ETS 08] should be respected. The first side lobe of the tag response signal should not exceed the effective radiated power (ERP) of −20 dBm at f_0, and nor should all the other spurious signals like its harmonics.

5.2.1.2. *Regulations at $3f_0$*

In Europe, the spectrum regulations allocate the frequency band of 2500–2690 MHz to broadband mobile services (e.g. Long-Term Evolution (LTE) and Worldwide Interoperability for Microwave Access (WIMAX) with transmitters of tens of watts and operation distances up to 4 *Km*) [ECC 12]. In other regions (e.g. where the harmonic frequency $3f_0$ = 2745 MHz), the band of 2700-2900 MHz is commonly allocated to meteorological and air traffic radar systems with high power (the equivalent isotropically radiated power could be in the order of 10^9 watts) [ECC 12, NTI 14]. Studies, based on worst case assumptions (i.e. line-of-sight conditions), have shown that there is potential interference from mobile services to radar and vice versa. This will depend on the deployed factors such as frequency separation, antenna orientation and distance (in the order of kilometers). Nevertheless, mobile and radar systems are still compatible [ECC 12].

Given the backscattering modulation nature of passive UHF RFID systems, even if the tag response at $3f_0$ is enhanced for the harmonic communication approach, it remains weak (the equivalent isotropically radiated power in the order of 10^{-11} watts at few meters) [AND 14] and does not represent an incompatibility issue compared with the possible radar effects over the mobile service. If reader and tag mask regulations are respected at f_0, the operation at $3f_0$ ensures the compliance with regulations.

5.2.2. *Harmonic reader considerations*

In a reader-to-HTs communication scenario, some considerations should be taken into account for the dimensioning of the harmonic branch receiver of an envisaged RFID harmonic reader. Specific details in the baseband treatment and the analog-to-digital conversion are out of the scope of this work, and can be taken as a future line of research. However, some directives on the harmonic RFID reader design can be stated. A critical issue in the design of traditional RFID readers (i.e. working at f_0) is the suppression of the carrier wave signal that unintentionally leaks into the received path due to the use of common isolation techniques as circulator or couplers. The reader has to transmit a strong carrier wave signal to supply passive tags with sufficient operating power. The response of the tag on the other hand is generated by passive backscattering, which is achieved by a modulation of the RFID chip input impedance. As only a small fraction of the power incident upon the tag is reradiated, the signal received by the interrogator antenna is very weak compared with the transmitted carrier wave signal [MAY 09]. Therefore, the reduction of the transmitted leak is crucial and leads traditional reader architectures that aim to increase the reception sensitivity of RFID readers [MAY 09, LAS 09, MAY 09].

From measurement results in passive UHF RFID [AND 13], the carrier wave in reception at $3f_0$ is around 60 dB below the carrier wave at f_0. Using harmonic backscattering, in case of a receiver at $3f_0$, the limitations imposed by the carrier wave leakage do not represent a great inconvenience. Indeed, the conditions are suitable to design a receiver with higher sensitivity than the traditional receivers at f_0. Some noise may result from $3f_0$ leakage due to the synchronization circuitry at reception but it is much smaller than the circulator or coupler leakage at f_0 [YUN 14]. As in traditional passive RFID, the overall system performance in the harmonic communication scenario is limited by the tag. Thus, exploiting the harmonic signal produced by RFID chips at $3f_0$ with the proper antenna design offers a gain in the tag-to-reader power link budget at $3f_0$.

5.2.3. *Harmonic tag design*

In the intention to integrate the antenna–chip impedance matching network directly in the antenna design, the RFID antenna design is widely documented (impedance matching, size-reduction techniques, etc.)

[MAR 08, RAO 05]. In the case of the HT, the main difference with respect to traditional tag antennas is the dual-band operation. As discussed in section 4.4, the HT should include a distributed matching network that cancels the RFID chip reactance at f_0 and $3f_0$. Depending on the envisaged application, some antenna parameters can be optimized with respect to others (size, directivity, materials, etc.). The HT design process is illustrated in a flow chart shown in Figure 5.1. As in traditional RFID [RAO 05], the design requirements are linked to the envisaged application (size, directivity, read range, operation conditions, materials of the tagged object, etc.).

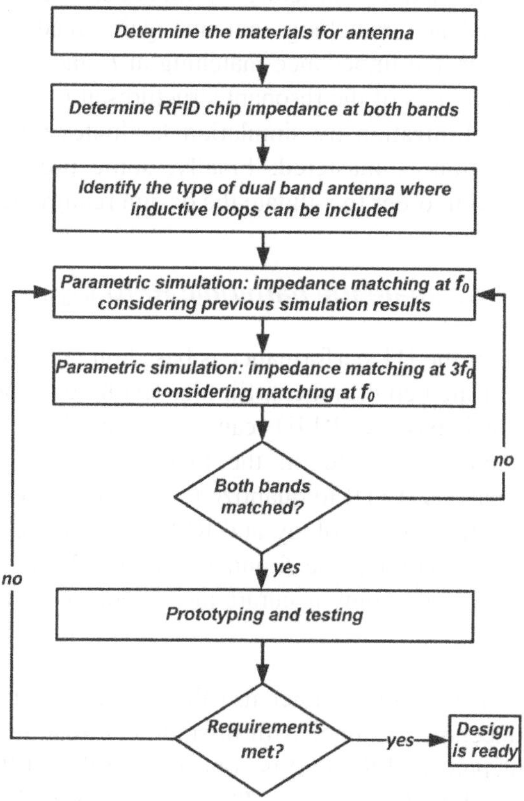

Figure 5.1. *RFID HT antenna design process*

Having fixed the material, the design of an HT antenna begins by importing the scattering parameters of the chips in the scavenging state into a 3D electromagnetic simulator [AND 14]. The impedance of the RFID chip at $3f_0$ is still capacitive [AND 14], and therefore techniques similar to the inductive loop matching used at f_0 can be applied.

Regarding the antenna structures, useful designs and techniques used for rectennas and harmonic radar can be applied in the design of the HT antennas [HEM 14, SCH 14, RAS 14, RAS 14] because of the common idea of inductive compensation of capacitive diode-based circuits. Under this criterion, the chosen antenna structure should include two inductive loops, each one to cancel the capacitive effects of the chip at f_0 and $3f_0$. The simulation is usually led by structure optimization based on the design requirements (impedance matching, antenna gain, read range, de-tuning effects). Starting with the impedance matching at f_0 and after at $3f_0$, certain changes in the dimensions of the resonant structures at f_0 could be necessary; nevertheless, the optimization by simulation facilitates the procedure once the structure operation is mastered. Finally, some prototypes should be fabricated and tested in order to validate the design requirements.

5.2.4. *Metrics to evaluate the harmonic RFID tags*

The evaluation of the HT performance is based on radiated measurements performed in a confined environment. Similar testing environments to those used for traditional passive RFID can be used [NIK 12], but some considerations should be made on the methods used, i.e. using power spectral density analysis (PSD) to quantize the HT response. Specific details about the measurement methodology and RFID air interface setting used to evaluate harmonic signals can be found in Chapter 3. A calibrated power budget should allow performing accurate measurements of the HT response at f_0 and at $3f_0$.

The tests reported in this chapter use the measurement setup shown in Figure 5.2. An anechoic chamber with bi-static configuration (1 m distance between horn antennas and the tag) is connected to the RFID Test-Platform (RFID-TP). The RFID-TP performs a PSD analysis in real time at f_0 and at $3f_0$. The results reported in this chapter consider the following directives regarding the power budget and measurements:

– HT sensitivity at different frequencies: the necessary ERP power sent by the RFID-TP to activate the HT is considered (i.e. transmitted power before the antenna). The ERP power is calculated using the power approach defined in [FED 15]. Since the measurement setup is completely calibrated and controlled, the ERP is known. This consideration allows us to evaluate the HT performance independently of the kind of reader antenna used;

– HT response at different RFID-TP powers: the HT response is evaluated in terms of the isotropic received power (i.e. values calibrated to a receiving antenna of 0 dBi gain) as a function of the Equivalent Isotropically Radiated Power (EIRP) transmitted by the RFID-TP. Furthermore, in the measure of the HT response, the PSD analysis considers the addition of power from both first side lobes of the modulated response, contrary to calculations explained in Chapter 3, where only one side lobe is considered. It is worth noting that the harmonic response exists conditioned to the existence of the response at f_0;

– HT read range performance: the read range is calculated using the law of decay $20log(d)$, where d is the distance between tag and reader beyond 1 m (1 m is defined by the calibrated setup in the anechoic chamber) and considering a radiated level of 35 dBm EIRP for the RFID-TP [ETS 08]. Therefore, the calculation of the HT response at f_0 and at $3f_0$ beyond 1 m is possible and it is compatible with the methods used in the literature [RAO 05]. For the read range evaluation, a 6 dBi antenna gain is considered for the EIRP of the RFID-TP. This analysis allows the representation of the read range for a given sensitivity of the reader receiver at f_0 and $3f_0$.

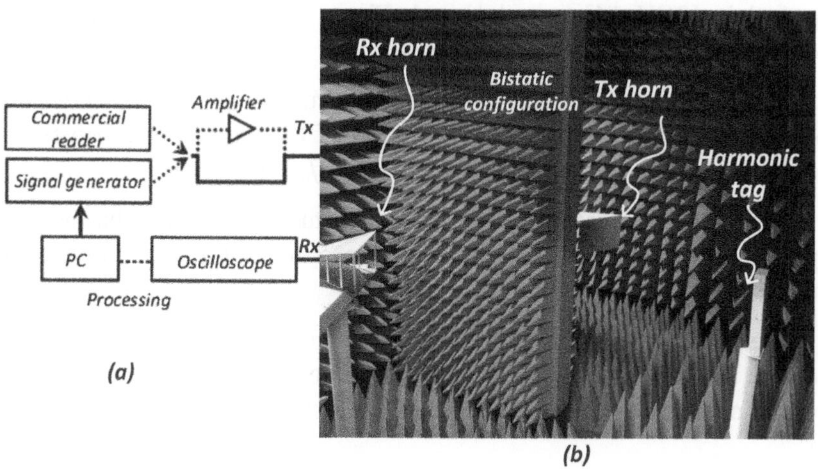

Figure 5.2. *a) Equipment setup for bi-static configuration; b) anechoic chamber configuration*

5.2.5. *Application case and experimental results: harmonic tag design example*

The requirements for the HT in this example are: (1) a maximum read range at f_0 = 868 MHz and f_0 = 2604 MHz; (2) an antenna size comparable to the dimensions of existing RFID antennas, and (3) the use of low-cost materials and prototype techniques. The materials chosen are polyethylene terephthalate (PET) with 3.2 permittivity and 0.05-mm thickness for the substrate as in traditional RFID tag inlays and aluminum as a trace conductor. Regarding the impedance at both bands, the design example uses two kinds of chips: (1) G2XM NXP SOT1040-AA1 [NXP 13] with measured impedances of 19–j178 Ω at f_0 and 9.2–j54.2 Ω at $3f_0$ and (2) a Monza chip with measured impedances of 20–j170 Ω at f_0 and 3.4–j29.4 Ω at $3f_0$ (conducted measured details in Chapter 3). The next step is to choose an antenna structure as reference where the double inductive loop matching technique can be applied. The following section pinpoints the design procedure for two kinds of antennas.

5.2.5.1. *Antenna design*

5.2.5.1.1. Yagi-Uda structures

A modified Yagi-Uda antenna following similar structural approaches as reported in [SCH 14] was designed. Figure 5.3 shows the Yagi-Uda HT prototype composed of three main parts:

– Fundamental resonator: the resonant structure at f_0 is a curved dipole conforming a loop (Figure 5.3(a)), that, at the same time, serves as an inductive loop for the impedance matching at f_0 [MAR 08];

– Harmonic resonator: the resonant structure at the third harmonic is a dipole joint to an inductive loop (Figure 5.3(b)), for the conjugate chip impedance matching at the third harmonic;

– Reflector: a reflector at the rear of the antenna (Figure 5.3(c)) serves to increase the directivity at the harmonic frequency; and, at the same time, its position and dimensions allow tuning the resistance of the antenna.

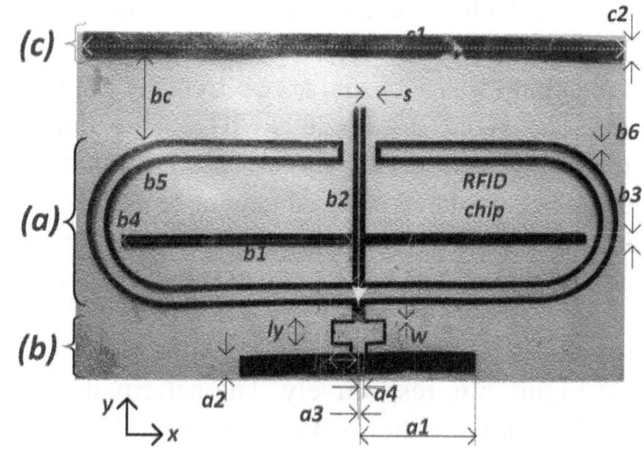

Figure 5.3. *HT prototype. a) Fundamental resonator; b) harmonic resonator; and c) reflector*

The harmonic resonator and the fundamental resonator are joined by two parallel lines, where in each line, one port of the chip is connected. The antenna can be tuned by trimming the dipoles and the reflector, the distance between the parallel lines *a3*, and the size of the inductive loops. Table 5.1 summarizes the optimum design parameters.

Parameter	Value (mm)	Parameter	Value (mm)
a1	18.6	b6	0.8
a2	3.3	c1	85
a3	0.4	c2	3.3
a4	0.8	lx	2.2
b1	35.5	ly	3.3
b2	27.2	w	0.4
b3	1.2	bc	12.8
b4	9.3	s	1.5
b5	11.9	-	-

Table 5.1. *Yagi-Uda antenna type without a director: design parameters*

The simulated reflection coefficient Γ of the harmonic antenna normalized to the chip impedance is depicted in Figure 5.4. The conjugate impedance chip matching was achieved at two frequencies, with Γ = −14.5 dB at 868 MHz, and Γ = −10.1 dB at 2604 MHz. This dual-band matching represents the main difficulty during the HT design process, and triggers in small bandwidths. The simulated realized gain at 868 MHz and at 2604 MHz is of 2.5 dBi and of 2.9 dBi, respectively. It is worth noting that different dual band antenna structures can be used [JIA 10, HUC 13].

The simulated directivity in the plane E (xy) and plane H (zy) is depicted in Figure 5.5 and Figure 5.6, respectively. The pattern at f_0 is comparatively closer to an omnidirectional pattern. By contrast, at the third harmonic, the pattern is slightly directive due to the antenna structure which improves the harmonic performance in one direction.

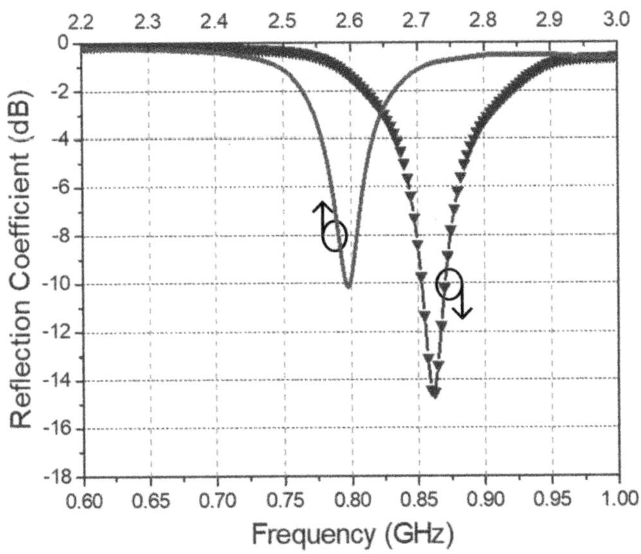

Figure 5.4. *Simulated reflection coefficient of the HT, normalized to the chip impedance*

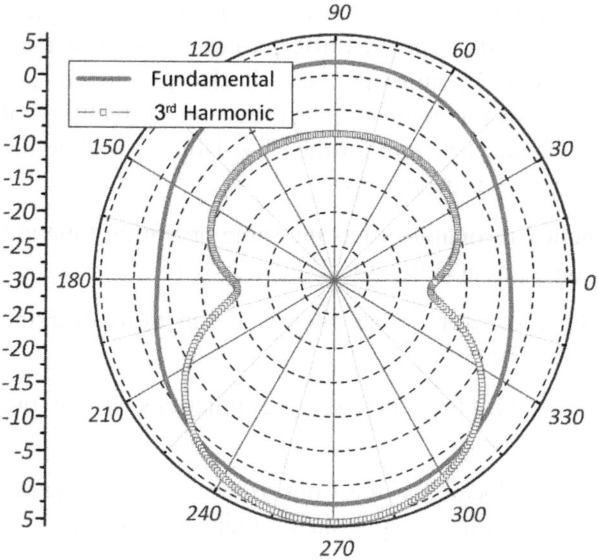

Figure 5.5. *Simulated directivity in the plane E for the HT*

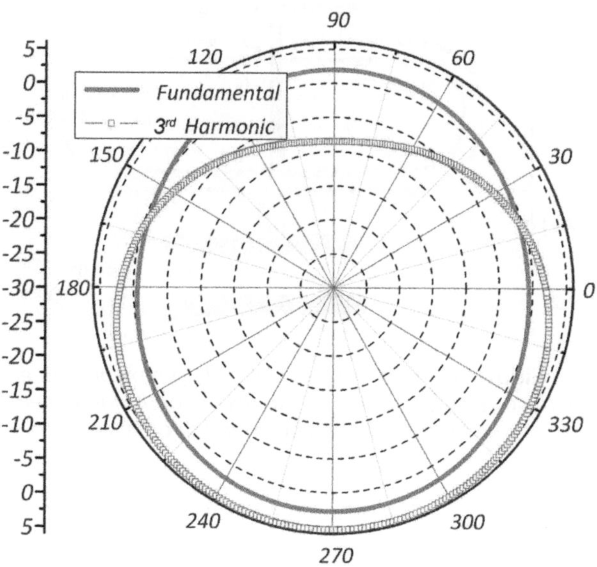

Figure 5.6. *Simulated directivity in the plane H for the HT*

In order to enhance the performance of the HT and profiting from the options that Yagi-Uda structures offer, a new version of the antenna above presented, and with the same materials, is also developed. This new version incorporates a director element to increase the directivity of the antenna. Figure 5.7 shows the structure for the Yagi-Uda HT with a director, which is now composed of four main parts:

– fundamental resonator: it has the same operation functions described in the HT version without a director (Figure 5.7(b));

– harmonic resonator: it has the same functions described in the first version (Figure 5.7(a));

– reflector: it has the same serving functions as the previous version (Figure 5.7(c));

– director: it is located at the front of the antenna (Figure 5.7(d)), which aims to increase the directivity at the harmonic frequency.

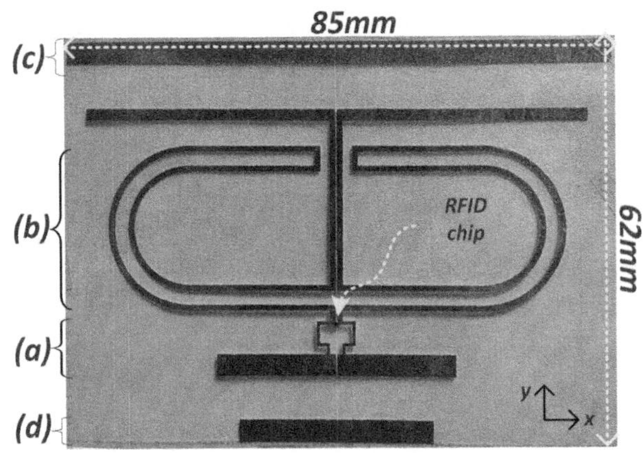

Figure 5.7. *New version of the HT prototype with a director element. a) Harmonic resonator; b) fundamental resonator; c) reflector and d) director*

The position and dimension of the director are optimized by simulation, with the same tuning policies as the HT version without a director, mentioned above, being followed for other structure elements. Table 5.2 summarizes all the optimum design parameters for the new HT antenna.

Parameter	Value (mm)	Parameter	Value (mm)
a1	18.6	c1	85.0
a2	3.3	c2	3.3
a3	0.4	lx	2.9
a4	0.9	ly	3.8
b1	38.1	w	0.4
b2	25.5	bc	12.8
b3	1.7	s	1.5
b4	8.8	d1	30.6
b5	11.9	d2	3.4
b6	0.9	–	–

Table 5.2. *Yagi-Uda antenna type with a director: design parameters*

With this structure arrangement, a hybrid directivity of the HT with a director is composed of:

– a quasi-omnidirectional pattern at f_0, i.e. the directivity at 868 MHz is now 2.18 dBi, slightly higher than the 1.8 dBi reported in the HT without a director;

– a directive pattern at $3f_0$, i.e. the directivity at 2604 MHz is 6.31 dBi, being also slightly higher than the 6 dBi reported in the HT without a director.

Readings at f_0 and/or the third harmonic frequency for the HT with hybrid directivity introduce an information data unit, because they consider the HT positioning with respect to the reader. Figure 5.8 and Figure 5.9 show the directivity of the HT with and without a director in the plane E (xy) and plane H (zy), respectively.

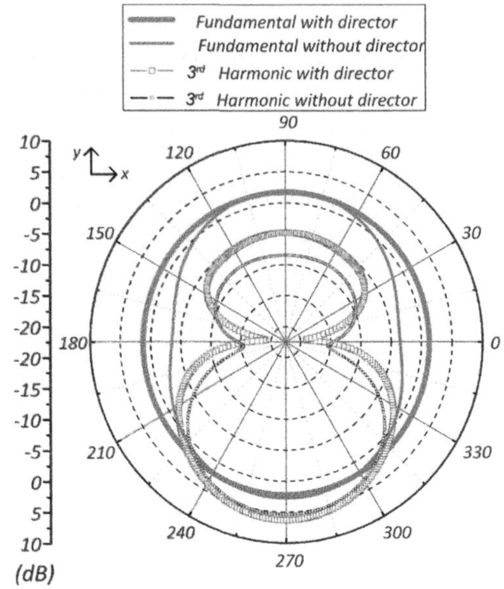

Figure 5.8. *Simulated directivity in the plane E for the Yagi-Uda HT with and without a director*

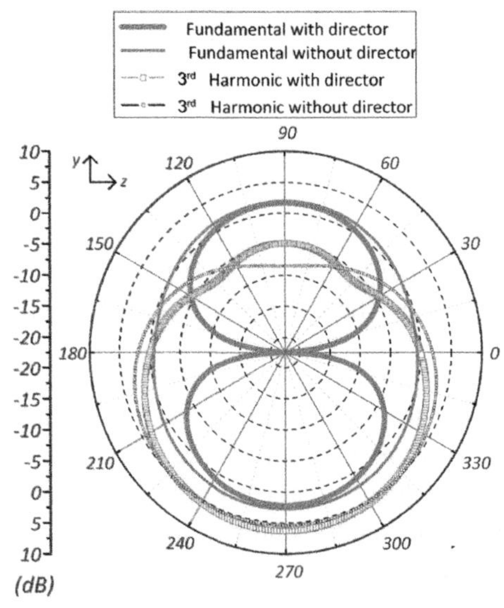

Figure 5.9. *Simulated directivity in the plane H for the Yagi-Uda HT with and without a director*

5.2.5.1.2. LF-inverted structures

A modified LF-inverted antenna is designed [JIA 10] on substrate PET with permittivity of 3.2 and 0.05 mm thickness. Each part of the planar inverted-F structure is designed and optimized to be able to resonate at the UHF band centered on 868 MHz. The inverted-L part of [JIA 10] becomes here a complete loop going back to the ground, each part of which, along with the distance between the inverted-F, is optimized to have a resonance at 2604 MHz. The capacitive impedances seen by the antenna at f_0 and at $3f_0$ are reduced by creating inductive loops in the ground. Figure 5.10 presents the structure of the antenna hereafter called HT2. Table 5.3 summarizes all the optimum design parameters for this structure.

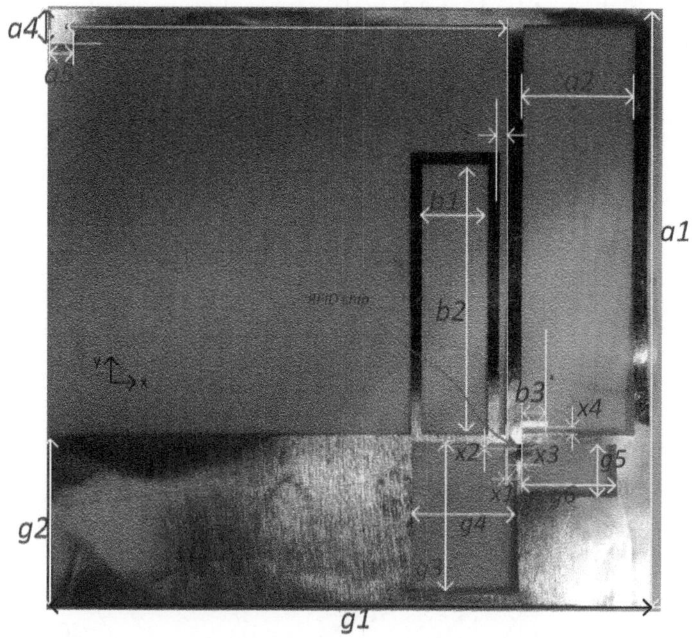

Figure 5.10. *HT prototype in the LF-inverted structure*

Results for directivity in the plane E (xy) and plane H (xz) are depicted in Figure 5.11 and Figure 5.12, respectively. With this structure arrangement, the HT is also composed of a quasi-omnidirectional pattern at f_0 and a directive pattern at $3f_0$, with 1.44 dBi and 6.01 dBi directivity, respectively.

Parameter	Value (mm)	Parameter	Value (mm)
g1	62.0	a5	2.3
g2	17.0	b1	7.0
g3	15.6	b2	28.0
g4	11.0	b3	2.7
g5	5.6	x1	0.2
g6	10.0	x2	0.2
a1	60.4	x3	0.2
a2	11.9	x4	1.0
a3	47.5	x5	1.0
a4	3.3		

Table 5.3. *LF-inverted antenna parameters*

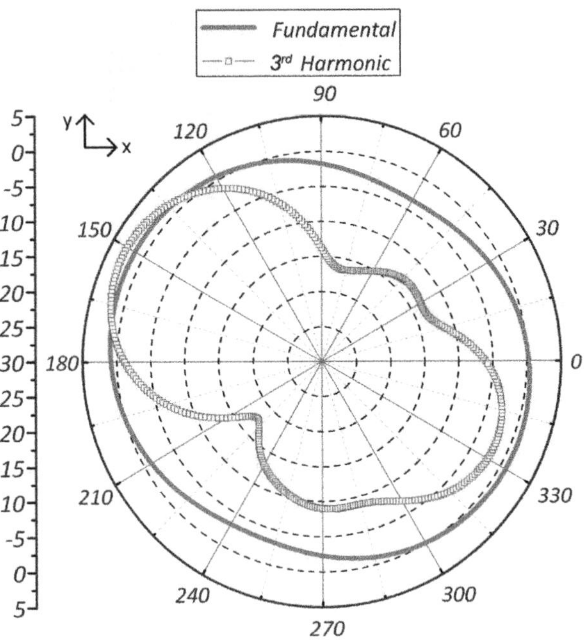

Figure 5.11. *Simulated directivity in the plane E for the LF-inverted HT*

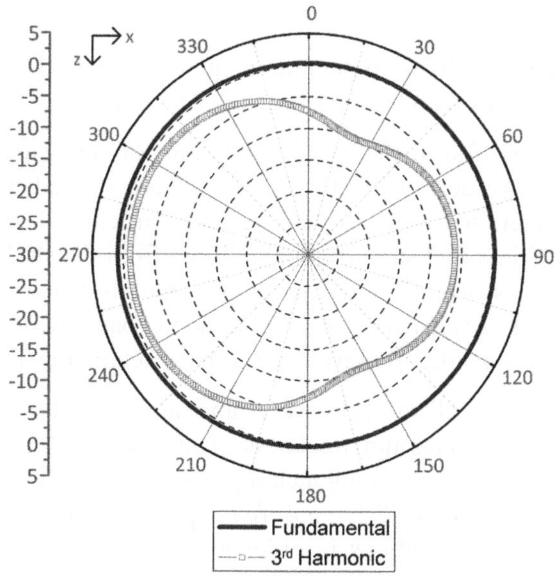

Figure 5.12. *Simulated directivity in the plane H for the LF-inverted HT*

5.2.5.1.3. Theoretical analysis of the HT sample

The HT with Yagi-Uda structure is chosen for this analysis. In order to estimate the Q factor of the distributed matching network included in the HT example, from [4.7] it is necessary to know R_{eq} and R_A for each HT. R_{eq} for each chip is obtained by converting the series RC impedance measured to a shunt RC equivalent model. Thus, an R_{eq} = 1.7 kΩ and an C_{eq} = 1.03 pF is obtained. The calculation of R_A is not straightforward, given that antenna structures already consider an integrated matching network, and its input impedance in a resistive design is unknown. Nevertheless, simulations were carried out to approximate the values of R_A. In the case of the HT example, the simulation mainly considers: (1) a fulfilled conductor for the fundamental resonator, shown in Figure 5.3(b), avoiding the internal inductive gap effect and (2) a feeding port at the middle of the fundamental resonator. An R_A = 61.7 Ω is obtained for the chosen antenna, allowing us to calculate a quality factor Q at f_0, equal to 5.2 and Q^* the one at $3f_0$, equal to 2.1 for the HT prototype. In agreement with the analysis presented in section 4.4, the HT example offers a similar treatment to the signals produced by the chip at f_0 and at $3f_0$.

5.2.5.2. *Experimental results*

In this section, results from parametric analysis on the HT prototypes are presented. The methods and setup are detailed as follows:

– HT under test: four different configurations of HT using the two kinds of antenna structure described in section 5.2.5 are studied. Henceforth, the notation HT1 refers to tags with the Yagi-Uda structure, and HT2 refers to tags with the LF-inverted structure. In addition, from the study performed in section 3.2, and as a comparative reference, the so-called commercial T5 tag with the best performance at the harmonic band was chosen for the study. The four HT configurations under test are detailed in Table 5.4.

– HT prototyping: all the HTs were fabricated on a PET substrate with 3.2 permittivity and 0.05 mm thickness. The common trace conductor is aluminum, and the chip mounting technique is flip-chip [RAO 00]. For each HT configuration shown in Table 5.4, five samples were fabricated with the same procedure and conditions in order to realize a fabrication sensitivity study. To be meaningful, only the most representative results will be reported.

– Measurement setup: the performance of the prototypes was evaluated in an anechoic chamber with the same RFID Test Platform (RFID-TP) and configuration used in section 3.2 (1 m distance between RFID-TP and tag). It is worth noting that the RFID-TP considers the communication protocol between reader and tag while tests are performed [EPC 15].

– The HTs under test were evaluated under a parametric analysis performed on three main parameters: (1) performance of the HTs at the UHF RFID band, the results of which are shown in section 5.2.5.2.1; (2) performance of the HTs at different powers sent by the RFID-TP, at f_0 and at $3f_0$, the results of which are shown in section 5.2.5.2.2; and (3) performance of the HTs at different reading distances, at f_0 and at $3f_0$, the results of which are shown in section 5.2.5.2.3.

– Results for each one of the studied parameters are organized considering four main features: (1) fabrication sensitivity, (2) bandwidth, power or read range, in each parameter case, (3) kind of chip used, and finally, (4) the performance of the HT compared with the commercial tag T5.

– All figures use the same legends to represent each HT configuration in a different type of line. The straight line represents the results for the tags of type HT1 (or HT2-M), the dashed line represents the results for the tags of type HT1-D (or HT2-N) and the symbol line represents the results for the tag T5. Different colors of the same type of line correspond to a different sample of the same HT configuration.

Antenna structure	**HT configuration label**	**RFID chip**	**Figure legend**	**Structure**
HT1 Yagi-Uda (without a director)	HT1-S	NXP UCODE [NXP 13]	Straight line	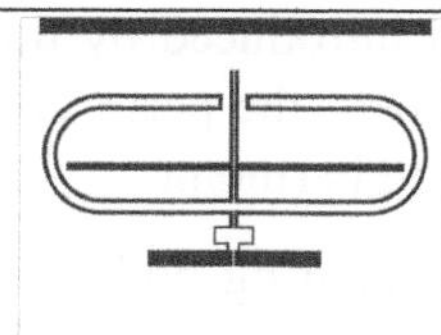
HT1 Yagi-Uda (with a director)	HT1-D	NXP-UCODE [NXP 13]	Dashed line	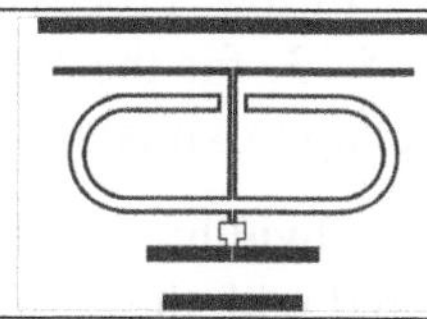
HT2 LF inverted	HT2-N	NXP-UCODE [NXP 13]	Straight line	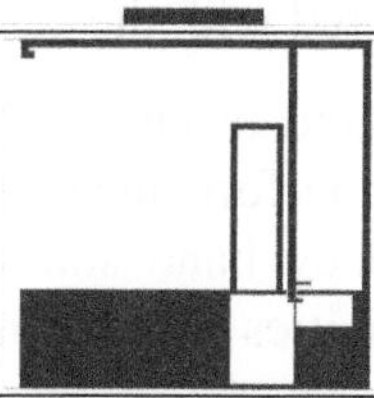
HT2 LF inverted	HT2-M	Monza 5 [IMP]	Dashed line	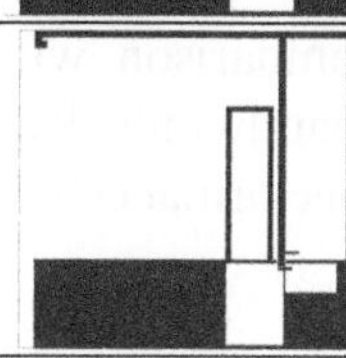
Dipole (commercial tag [UPM 13a, UPM 13b, UPM 13c, UPM 13d])	T5	–	Symbol line	

Table 5.4. *Different configurations of the HTs under test*

5.2.5.2.1. Performance of the HTs in the UHF RFID band

Following the nomenclature presented in Table 5.4, Figure 5.13 and Figure 5.14 show the Effective Radiated Power (ERP) transmitted by the RFID-TP, as a function of the frequency, respectively, for the Yagi-Uda structure and the LF-inverted structure. Each HT sensitivity is evaluated in terms of the ERP (transmitted power before the antenna). The ERP evaluation allows us to perform a general analysis, independent of the kind of reader antenna used. The results of the analysis are summarized as follows:

– Fabrication sensitivity: design directives for all HTs imposed a first resonance centered in 868 MHz. From Figure 5.13 and Figure 5.14, in both structure cases, a de-tuning effect is observed, mainly due to the capacitive effect introduced by the flip-chip mounting process. Despite the evidenced de-tuning, the performance of the HTs is always evaluated at 868 MHz and its third harmonic.

Comparing both kinds of structures, HT1 tags are most sensible to the manufacturing process due to the smaller dimensions. As shown in Figure 5.13, for the same structure (HT1 or HT1-D) and the same kind of chip, an approximately 15 MHz variation in the resonance can be observed.

– Bandwidth: with 20 MHz, the narrower bandwidth is for HT1 tags, compared with the 65 MHz bandwidth for HT2 tags.

– Kind of chip: tags with a Monza 5 chip (HT2-M) in Figure 5.14 have better performance in terms of the lowest activation power in the whole frequency band studied. These results are a consequence of the different sensitivities reported in data sheets for both kinds of studied chips.

– Comparison with commercial tag: the commercial tag T5 reports a wider bandwidth. Regarding the activation sensitivity, HT2-M tags have a better performance than T5.

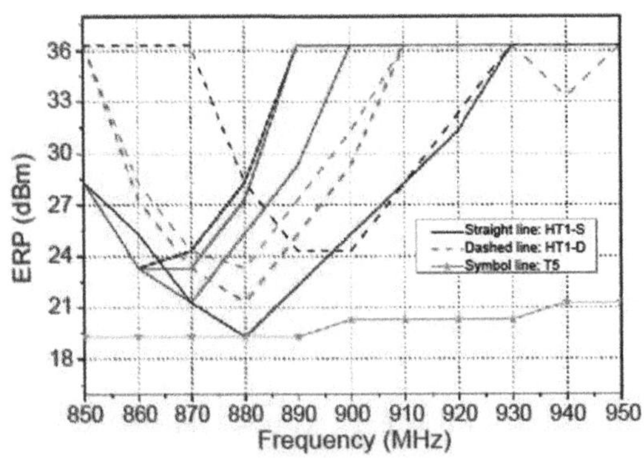

Figure 5.13. *ERP transmitted by the RFID-TP as a function of the frequency for the tags of type HT1 compared with a commercial tag. The analysis is performed at f_0*

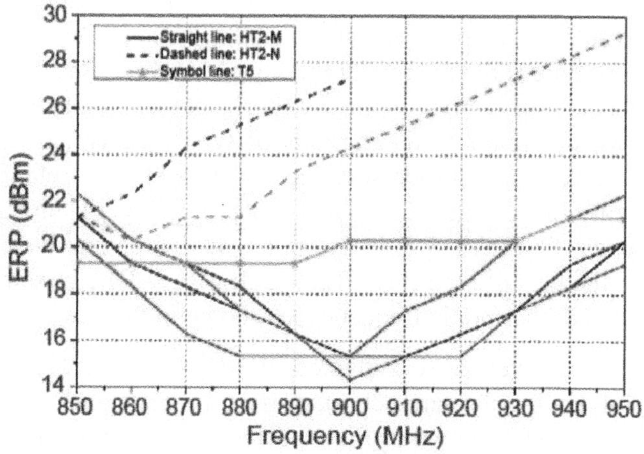

Figure 5.14. *ERP transmitted by the RFID-TP as a function of the frequency for the tags of type HT2 compared with a commercial tag. The analysis is performed at* f_0

5.2.5.2.2. Performance of the HTs at different powers transmitted by the RFID-TP

The HT response is evaluated in terms of the isotropic received power (values calibrated to a receiving antenna of 0 dBi gain) as a function of the EIRP transmitted by the RFID-TP. Furthermore, in the measure of the HT response, the power spectral density analysis considers the addition of power from both first side lobes of the modulated response, contrary to results reported in Chapter 3, where only one side lobe is considered.

Figures 5.15 and 5.16 show the results at f_0 for the tag configurations of type HT1 and tag configurations of type HT2, respectively. Figures 5.17 and 5.18 show the same results at $3f_0$. It is important to note that the harmonic response exists conditioned to the existence of the response at f_0, as a non-linear phenomenon of the RFID chip. The results of the analysis are summarized as follows:

– Fabrication sensitivity: considering the analysis frequency at 868 MHz and its third harmonic, the effects due to the manufacturing process in the

HTs performance is analyzed. A direct effect on the activation power that relies on the antenna–chip impedance matching is reported. From Figure 5.15, the tags of type HT1 (having all the same chip) are more sensitive with a 5 dB difference in the activation power (horizontal axis) than those of type HT2, as shown in Figure 5.16. The tags of type HT2 are in general less sensitive to manufacture processes due to straight and relatively large structure details. Indeed, the distributed matching network dimensions for the tags of type HT1 (Table 5.1: lx, ly, w, a3, a4, b4, b5 and b6) are shorter than those for the tags of type HT2 (Table 5.3: g1-g6). The same behavior is repeated at the third harmonic frequency, as shown in Figures 5.17 and 5.18;

– Tag response level: once activated, the tags of type HT1, shown in Figure 5.15 (all having the same chip), present a 5 dB higher response level than those of type HT2, shown in Figure 5.16. The reason is the higher gain of Yagi-Uda designs. Among the tags of type HT1, as expected, HT1-D tags have a 2 dB higher response level than HT1-S tags, due to the director element. The behavior is also repeated at the harmonic frequency, shown in Figure 5.17, reaching in the best case a third harmonic response of −77 dBm for HT1-D. Lower harmonic responses are shown for the tags of type HT2 in Figure 5.18.

– Kind of chip: besides the lower activation power of Monza 5 chips, the kind of chip also influences the hysteresis of the tag response. Indeed, for NXP-UCODE chips, the tag response remains quasi-constant upon activation even if the reader power increases as can be seen for HT1 and HT1-D in Figure 5.15 and HT2-N in Figure 5.16. On the contrary, for Monza 5 chips, the tag response increases with the reader EIRP, reaching a constant or saturated level at higher powers. A similar behavior is also observed at the third harmonic frequency in Figures 5.17 and 5.18. This behavior is intrinsically related to the internal (and proprietary) circuit architecture of each manufacturer, specifically the voltage regulator design [DEV 05].

– Comparison with commercial tag T5: in all cases, the proposed HTs have better performance (lower activation power and higher tag response) than the commercial tag T5, at f_0 and $3f_0$.

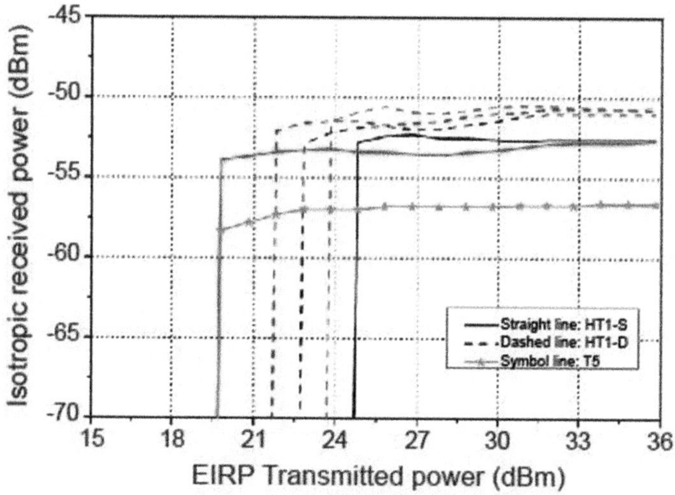

Figure 5.15. *Power of the tag response from the tags of type HT1 at f_0 as a function of the EIRP transmitted by the RFID-TP. The measured power considers an isotropic antenna at reception*

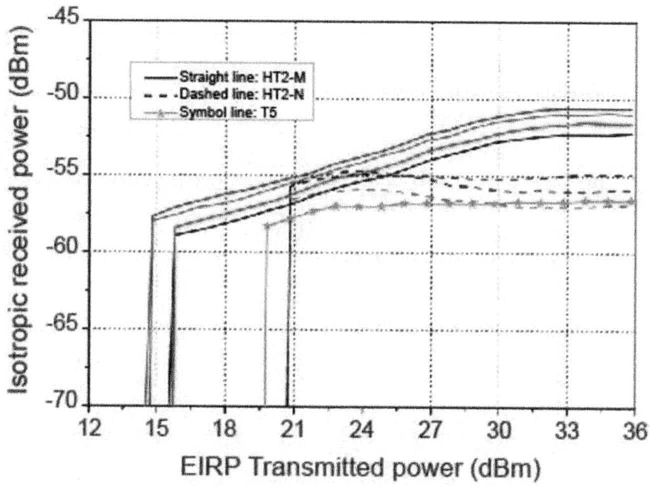

Figure 5.16. *Power of the tag response from the tags of type HT2 at f_0 as a function of the EIRP transmitted by the RFID-TP. The measured power considers an isotropic antenna at reception*

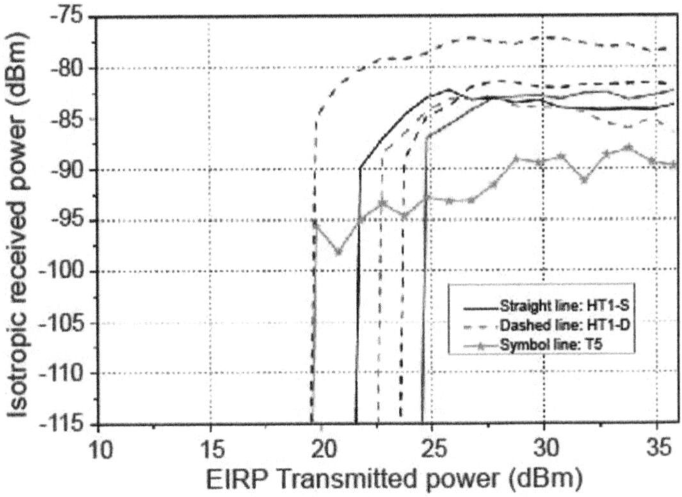

Figure 5.17. *Power of the tag response from the tags of type HT1 at 3f$_0$ as a function of the EIRP transmitted by the RFID-TP. The measured power considers an isotropic antenna at reception*

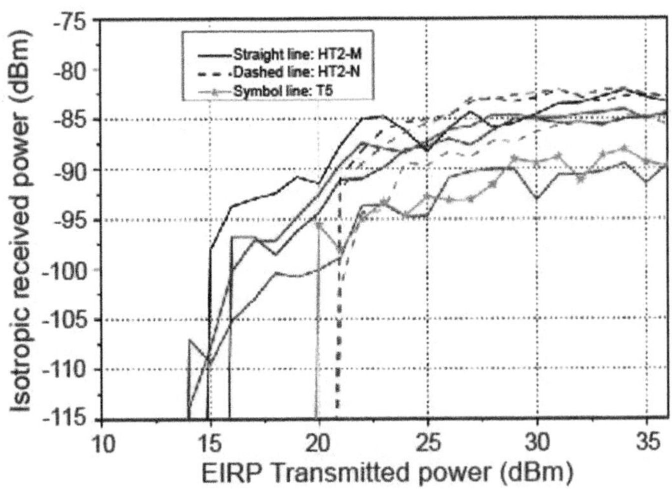

Figure 5.18. *Power of the tag response from the tags of type HT2 at 3f$_0$ as a function of the EIRP transmitted by the RFID-TP. The measured power considers an isotropic antenna at reception*

5.2.5.2.3. Parametric study on read range

Using the passive RFID read range equation [RAO 05] and considering the measured sensitivity of each HT, the tag response level at the receiver side is measured at f_0 and at $3f_0$. These measures consider a power transmitted by the RFID-TP (EIRP) of 35 dBm [ETS 08]. This analysis allows the representation of the tag read range for a given sensitivity of the receiver.

Figures 5.19 and 5.20 show the results at f_0, for the tags of type HT1 and type HT2, respectively. Figures 5.21 and 5.22 show the same results for the third harmonic frequency. The results of the analysis are summarized as follows:

– Fabrication sensitivity: the manufacturing process introduces an effect on the read range, which is different on each kind of HT configuration. From Figure 5.19, as expected, the tags of type HT1 are the most sensitive with a difference of 5.5 m read range (horizontal axis). The tags of type HT2 are in general less sensitive to this parameter. A similar behavior is observed at the third harmonic frequency in Figures 5.21 and 5.22.

– Read range: when a simultaneous operation is aimed, the third harmonic response limits the read range of HTs at f_0. For the case of HT1 tags in Figure 5.21, a read range larger than 5 m at the harmonic frequency is achieved if a receiver sensitivity of −90 dBm is attained [WAN 07]. For the case of HT2 tags in Figure 5.22, a read range of nearly 4 m at $3f_0$ is achieved with the receiver sensitivity of −90 dBm.

– Kind of chip: the chip sensitivity favors the read range at f_0, reaching for the Monza 5 chip a maximum of 9 m (HT2-M in Figure 5.20) if the receiver sensitivity is −70 dBm. Contrary to the harmonic frequency, the hysteresis effect of NXP-UCODE chips triggers at higher harmonic response levels, which favors the feasibility of the receiver sensitivity.

– Comparison with commercial tags: higher read ranges at f_0 and its third harmonic were achieved by the proposed HTs compared with tag T5. From Figure 5.21 or 5.22, if the receiver sensitivity was −90 dBm, barely 2 m read range is calculated for the commercial tag T5.

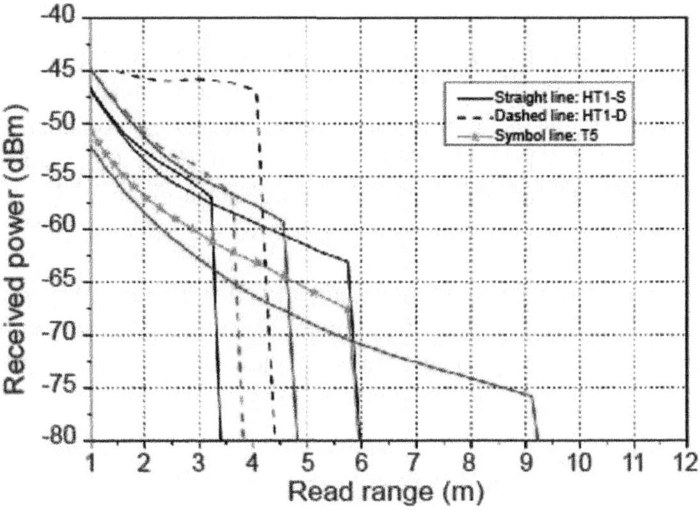

Figure 5.19. *Read range for the tags of type HT1 at f_0 as a function of 35 dBm EIRP transmitted by the RFID-TP. The measured power considers a 6 dBi antenna at reception*

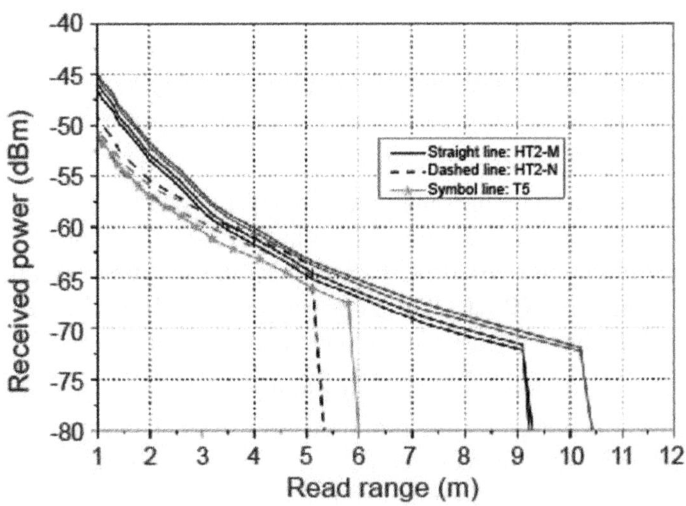

Figure 5.20. *Read range for the tags of type HT2 at f_0 as a function of 35 dBm EIRP transmitted by the RFID-TP. The measured power considers a 6 dBi antenna at reception*

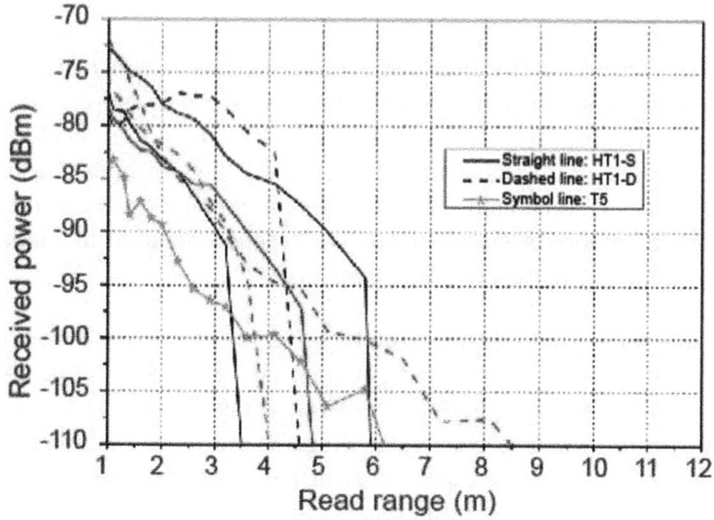

Figure 5.21. *Read range for the tags of type HT1 at the third harmonic frequency as a function of 35 dBm EIRP transmitted by the RFID-TP. The measured power considers a 6 dBi antenna at reception*

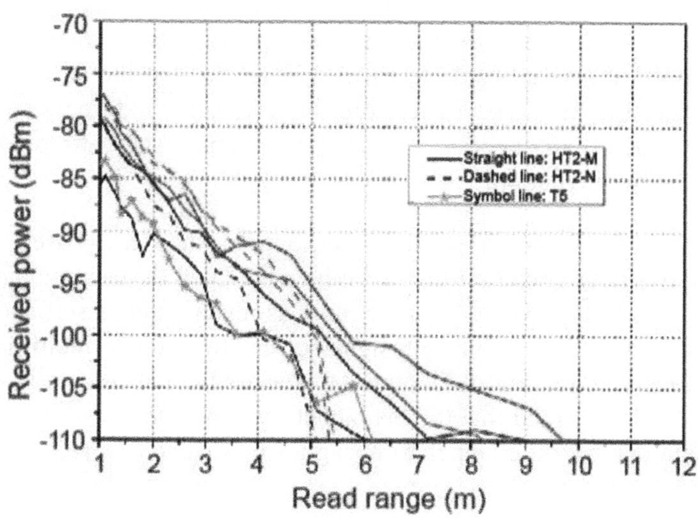

Figure 5.22. *Read range for the tags of type HT2 at the third harmonic frequency as a function of 35 dBm EIRP transmitted by the RFID-TP. The measured power considers a 6 dBi antenna at reception*

5.2.6. *Summary: harmonic tag*

Operation concepts for harmonic communication considering current regulations, harmonic readers, design guidelines for HTs and methods for its evaluations are discussed in section 5.2.5. The work on exploiting the response at the third harmonic by specific tag antenna design was experimentally demonstrated using HT design examples. The evaluation of the prototypes reports a simultaneous read range operation at fundamental and third harmonic frequency larger than 5 m.

The parametric study on the prototypes shows that the HT performance can be enhanced with the antenna structure and also the kind of chip. The frequency analysis shows that the LF-inverted tags have a wider frequency bandwidth than Yagi-Uda tags. Regarding the harmonic response strength, tags with the Yagi-Uda structure present a higher response of 5 dB than those with the LF-inverted structure. The kind of chip used is clearly an advantage in the HT performance enhancement. The chip sensitivity suits the read range at the fundamental frequency, reaching in the suitable case 9 m for Monza 5 chips, with a reader sensitivity of −70 dBm. Contrary to the harmonic frequency, the hysteresis effect of NXP UCODE chips suits the feasibility of the reader sensitivity in a third harmonic reception scenery because of the clear, higher and constant harmonic response produced by this kind of chip. The HTs present higher sensitivity to the fabrication process due to the significant electrical dimensions that the structures take around the third harmonic. Therefore, special care and mastery of the sources of losses are necessary in the fabrication stage to predict the performance by simulation.

Regarding the benefits of the contribution, they are based on the frequency diversity operation that uses the modulated third harmonic signal produced by the chip, where the spectrum has much reduced self and multi-path interference. The redundant signal offers to counter multi-path and leakage effects at reception. Contrary to the traditional efforts in passive RFID, it should be noted that the main contribution of this work is to have a second communication channel from tag-to-reader and not to increase the read range.

The next section will focus on a different approach for the exploitation of the deeply studied harmonics: electromagnetic harmonic energy harvesting.

5.3. Harmonic harvesting: empowering the RFID tag

In this section, the second proposition of augmented tags is presented. We introduce a new concept enabling the association of a passive UHF RFID tag with a commercial sensor, and this, without use of additional energy source. The idea relies on the introduced paradigm of the exploitation of the harmonics (namely, the third harmonic) produced by passive RFID chips to provide frequency diversity, as depicted in section 5.2. Here, a new method for exploiting the third harmonic is proposed. It involves harvesting the power carried by these commonly wasted harmonic signals to later supply power to a sensor.

This section recalls some theoretical foundations concerning the harmonic exploitation in passive RFID and describes the proposed concept of harmonic harvesting. The design methodology of the RFID tag harmonic harvester, which is based on the study of a non-linear equivalent circuit of the RF front-end of passive RFID chips, is detailed. Special attention is dedicated to the matching network design. The experimental validation of the proposed RFID chip model and the standard metrics to evaluate the harvesting performance of the designed circuits are presented. Finally, in order to validate the feasibility and to evaluate the performance of the augmented tag, an application example in which the energy harvested from the third harmonic of a RFID chip is used to supply power to a commercial temperature sensor.

5.3.1. *Harmonic generation in diode-based circuits*

Most of the energy-harvesting circuits contain a low-pass filter (LPF) at the input that allows the fundamental frequency to enter, and rejects the harmonics back into the energy harvester. This technique, very common in high-efficiency energy harvester design, allows the maximization of the conversion efficiency by confining all the higher order harmonics between the LPF and the DC pass filter. Harmonic treatment techniques such as the Class-E or Class-F rectifiers [GEO 10] help to confine the energy carried by harmonics in order to optimize the conversion efficiency of rectifier circuits.

The techniques are based on presenting impedances to the rectifier circuit (at fundamental and harmonic frequencies), enabling the stored energy in the diode not to be self-dissipated, but rather released to the load, and therefore, enhancing its RF-to-DC conversion efficiency. The harmonic treatment can be addressed in a design stage by simulation tools such as the harmonic balance (HB) [GEO 10]. These efficient design techniques can also be used in passive RFID tags if antenna and RFID chip are co-designed and optimized together as a single device as in the case of efficient rectennas [YON 04].

However, for industrial interests and practicality of RFID labels manufacturing, RFID chips and RFID tag antennas are separately designed and optimized only at the regional frequency band of operation (e.g. 868-870 MHz in Europe or 902-928 MHz in US, 950-956 MHz in some places in Asia). Therefore, and in agreement with the theory and measurements described in Chapter 3 and Chapter 4, passive RFID chips generate harmonics and some RFID tag antennas can radiate these harmonics.

5.3.2. *Techniques to empower the RFID chip and rectifier circuits in general*

A state of the art of the techniques to enhance the rectification efficiency of electromagnetic energy harvesters is reviewed below:

– Energy sources and harmonics: with the deployment of wireless sensor networks that become ubiquitous, an essential challenge is the energy autonomy of sensors. In parallel, in the field of passive RFID where tags are wirelessly powered by the reader, the transformation of tags in sensor tags (notably tags with embedded sensors) also requires new energy needs. In addition, the use of ecofriendly solutions is nowadays a great and strong requirement for all electronics design. Besides energy-harvesting solutions exploiting sources such as solar energy, human body temperature or vibration, two types of approaches based on technically similar electromagnetic (EM) sources are conceptually opposed: exploitation of ambient EM sources such as Wi-Fi or digital TV signals, often referred to as EM energy harvesting (EEH), versus the use of intentional signals as an energy source, the so-called wireless power transfer (WPT), whose RFID can be seen as a particular case.

– Harvesting harmonic power: a large number of investigations are being focused on wireless power or remote-powering technologies. Therefore, WPT has become a refreshed and refocused technique for powering electronic devices over distances. A rectenna circuit including antenna and rectifier is the most vital component of a microwave power transmission system that wirelessly receives microwave power and converts it into DC. Rectenna efficiency is mainly determined by its rectifier circuit. Hence, designing a high-performance rectifier is a point of interest since it is directly related to the quality of a WPT system. Few works are focused on the utilization of harmonic signals with energy-harvesting intentions. A millimeter-wave rectenna with high-efficiency rectifier operation for medium and low levels of RF input power has been proposed in [LAD 14]. The second harmonic signal generated by conventional rectifiers operating at medium range of RF input power is exploited by a modified voltage doubler rectifier that is able to harvest the energy from harmonics. The DC outputs of the harmonic harvester and the fundamental harvester are added up at the end in a common DC load.

– Signal design for WPT: several works have proposed the use of specific signal in a WPT in order to increase the RF-to-DC conversion efficiency of rectifying circuits. The basis behind these phenomena is the non-linear behavior of rectifying diodes under different input signals. Indeed, using specially designed signal waveforms at the transmitting terminal (that somehow excite the diode non-linear behavior in a more efficient way), the RF-to-DC conversion efficiency of the diode is increased.

The phenomena can be generalized by the use of time varying envelope signals as an efficient WPT technique, since they are capable of activating the rectifying devices for lower average input power levels, compared with signals of constant envelope and the same average power. Therefore, signals with large peak-to-average power ratios (PAPRs) provide higher RF-to-DC conversion efficiency than single carrier signals, meaning that for the same targeted distance, we need to transmit less power.

The PAPR (dB) of a waveform can be defined as

$$PAPR = 10log_{10}\frac{max\left(x^2(t)\right)}{\langle x^2(t)\rangle} \quad [5.1]$$

where $x(t)$ is the time domain waveform of the signal of interest and the symbol "< >" refers to the time average operator. In [COL 14], an interesting study regarding the PAPR of three different signals compared with a single tone signal was presented. Each signal was used as an input for a rectifier circuit optimized at 433 MHz and the results indicate the mentioned phenomenon: the higher the PAPR is in the input signal, the higher is the rectifier RF-to-DC conversion efficiency. Most of the reported results in the literature consider signal levels above –30dBm. However, this behavior might be power dependent on a threshold power, for instance, with very ultra-low power (below –30 dBm), the PAPR effect might not be significant for the rectifier efficiency.

The signal design for efficient WPT shows that even though high PAPR signals are not desired in communication systems due to the distortion that these signals suffer when amplified in the transmitter, in EEH systems high PAPR signals are desired for improving the RF-to-DC conversion efficiency of rectifiers.

5.3.3. *Third harmonic exploitation in passive RFID*

The non-linear i-v characteristic of a diode-based circuit imposes a current signal to be composed by fundamental and harmonic components as a function of the voltage. Three milestones summarize the theoretical analysis of a diode-based RF front-end of passive RFID chips:

– the analysis of the current produced by passive RFID chips including the fundamental and the harmonic components as a function of the signal level coming from the reader shows that the current magnitudes of fundamental and third harmonic frequencies (so-called f_0 and $3f_0$, respectively) become comparable when the input power increases;

– based on the inductive coupling matching technique, traditional matching networks included in tag antennas can attenuate the third harmonic by a factor of 17 dB to 32 dB, depending on the antenna quality factor. In some cases, the attenuated third harmonic component is backscattered and can be detected by the reader, as discussed in Chapter 3;

– the design of dedicated matching networks produces an increase in the third harmonic backscattering level allowing the design of an RFID tag with frequency diversity, i.e. a tag that generates a redundant signal for tag-to-reader in addition to the traditional one at the fundamental frequency. Considering an ideal case, the backscattered level only depends on the harmonic production feature of the used chip. Experimental read ranges of 4.5 m have been reported in section 5.2 for the HT operating at both frequencies.

Instead of re-radiating the information carried by $3f_0$ as presented in the previous section but considering the same theoretical foundations, this section proposes a new approach to exploiting $3f_0$ by means of harvesting its carried energy to supply power to attached low-consumption electronics.

5.3.3.1. *Optimal integration of harvesting sections*

The system architecture of the augmented tag is depicted in Figure 5.23. It is mainly composed of: (i) the RFID section operating at f_0 responsible for the RFID communication in compliance with the EPC Class1 Gen2 standard; (ii) the harmonic harvesting (HH) section operating at $3f_0$ in charge of the power harvesting of the wasted $3f_0$; (iii) a three-port matching network (TPMN) that aims to efficiently redirect the signals at f_0 and $3f_0$ between the RFID section, HH section and the tag antenna.

With similar principles to those used for the design of Class-E or Class-F rectifiers, the impedances presented at the RFID section and EEH section inputs at fundamental and harmonic frequencies are determined by the TPMN operation. The design of the TPMN can be driven by optimization tasks thanks to the spectral analysis performed in an electric simulator as presented in the next section. The design of the augmented tag aims to achieve an optimal performance in terms of wireless communication for the RFID section (i.e. maximizing the read range) and DC power generation in the HH section (i.e. maximizing the RF-to-DC conversion efficiency). It is worth noting that the design of the HH section must be led by the cohabitation with the RFID section and its exploitation, considering that the performance at a given power level may be different for each section.

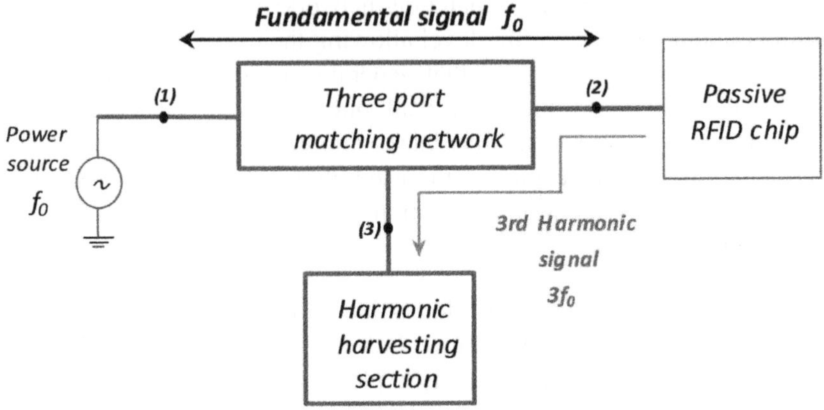

Figure 5.23. *Proposed RFID tag harmonic harvester, the so-called augmented tag*

In order to efficiently integrate the harmonic harvester circuit to a passive UHF RFID chip, this section details the design methodology of the augmented tag in four sequential steps: (i) the first step is the analysis and design of the RFID section based on an equivalent RF front-end of an RFID chip; the equivalent model allows us to accurately emulate the non-linear operation of the chip by simulation and to efficiently design the whole augmented tag; (ii) the second step is the design of the HH section considering an estimated input power, i.e. that of $3f_0$; (iii) the third step is the design of the TPMN considering the non-linearity of both the sections, RFID and HH; and (iv) the last step is the integration of RFID and HH sections with the TPMN to form the augmented tag depicted in Figure 5.23. The RFID reader is modeled by an RF source at 868 MHz (corresponding to the European band). ADS Keysight tools are used for all simulation studies.

5.3.3.1.1. RFID chip model

The RF front-end responsible for the non-linear behavior of a passive RFID chip, i.e. its rectifier part, is modeled in order to ensure a coherent behavior compared with commercial passive chips. The model is driven by the emulation of the input impedance in a wide frequency range including f_0 and $3f_0$, in order to describe its non-linear features such as harmonic generation and impedance power dependency. This feature leads us to consider a specific input power from which the design is optimized [GEO 10, YON 04]. The input impedance of the EM43235 RFID chip in the scavenging state is measured with an input power of 0 dBm using similar methods to those presented in section 3.3. This input power ensures a $3f_0$ product exhibiting a comparable level, in logarithmic scale, to the one reflected at f_0 (see section 3.2), besides being enough to activate the RFID chip. The EM43235 chip is chosen because of its size (TSSOP8 package) that offers an easy manipulation in a hybrid modular system.

Considering datasheet and Spice parameters for all components, a diode-based voltage doubler with four stages constituting a rectifier circuit with a configuration like the one used in the Dickson charge pump to design RFID chips [DEV 05] is used to emulate the RFID chip behavior. Figure 5.24 compares the input impedance of the measured chip with the one of the chip model showing an acceptable coherence.

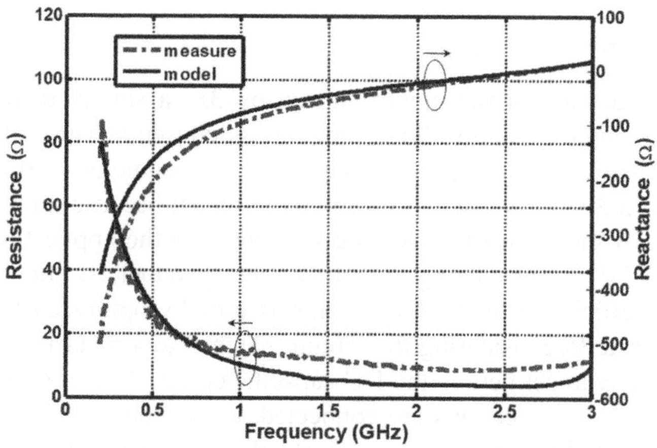

Figure 5.24. *Input impedance of the measured chip (EM4325) and the associated proposed model*

The optimization of the RF-to-DC conversion efficiency of the RFID chip model is performed for a 0 dBm input power at f_0, considering the circuit matched to a 50 Ω RF source. At this stage, all the lumped components of the chip model are defined. A final optimization is performed when all sections are connected; as it will be explained in section 5.3.3.2, this last optimization does not change the defined RFID chip model. The topology and component values of the passive RFID chip model are reported in detail in Figure 5.25.

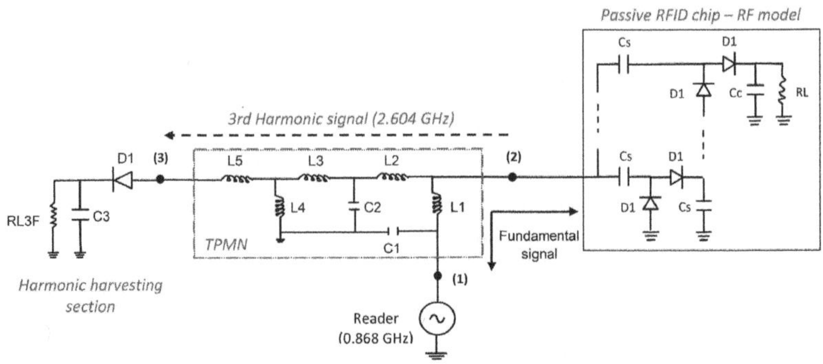

Figure 5.25. *Proposed harmonic harvesting system including a passive RFID chip model with Cs= 0.9 pF, Cc = 5 pF, RL = 7kΩ, D1 (HSMS285-C Schottky diode); a TPMN with C1 = 1.6 pF, L1 = 31.5 nH, L2 = 43 nH, C2 = 0.125 pF, L3 = 43 nH, L4 = 11.4 nH, L5 = 5.25 nH; and a harvesting section with D1, C3 = 2pF, RL3F = 3.3 kΩ*

5.3.3.1.2. Harmonic harvester section

In order to harvest the power carried by $3f_0$, a simple rectifier circuit based on a single Schottky diode topology is chosen because of its higher conversion efficiency compared with multistage topologies [BOA 13]. An HH load of 3.3 kΩ is chosen as design requirement because it represents the load value of the sensor that will be connected in the application example (section 5.3.4). Considering manufacturer parameters, the RF-to-DC conversion efficiency of the HH section is initially optimized to –10 dBm input power at $3f_0$ considering the circuit matched to a 50 Ω RF source. The only parameter to set here is the DC capacitance C3. A final optimization is performed when all sections are connected, as will be explained in section 5.3.3.2. The topology and the component values are reported in Figure 5.25.

5.3.3.1.3. Three-port matching network – TPMN

The TPMN shown in Figure 5.26 has three functions: (i) it matches the RFID chip impedance (port 2) to a 50 Ω input at f_0 (port 1) ensuring the power transfer from the antenna towards the RFID chip; (ii) it redirects the $3f_0$ from the RFID chip (port 1) towards the HH section (port 3); and (iii) it isolates the antenna input (port 2) from the HH section (port 3) at f_0.

Lumped TPMN

The non-linear impedances of the chip model and the HH section are first determined by large-signal S-parameter (LSSP) simulation. The chip model impedance is then matched to the 50 Ω input (RF source) at 0.868 GHz through L'1 and C'1. Then, the RFID section using the chip model and the HH section impedances are matched at 2.604 GHz by L'2, C'2 and L'3. This configuration triggers in the preliminary matching network shown in Figure 5.26(a). This first configuration does not ensure good f_0 isolation at the HH section, and could decrease the activation threshold of the RFID chip. Therefore, a second structure including inductors L4 and L5 is designed to overcome this problem. A final LSSP optimization using this configuration is run to obtain the final values of the TPMN shown in Figure 5.26(b).

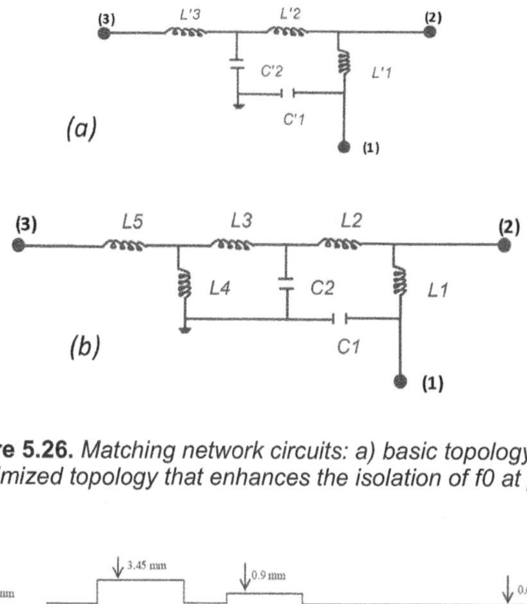

Figure 5.26. *Matching network circuits: a) basic topology and b) optimized topology that enhances the isolation of f0 at port 3*

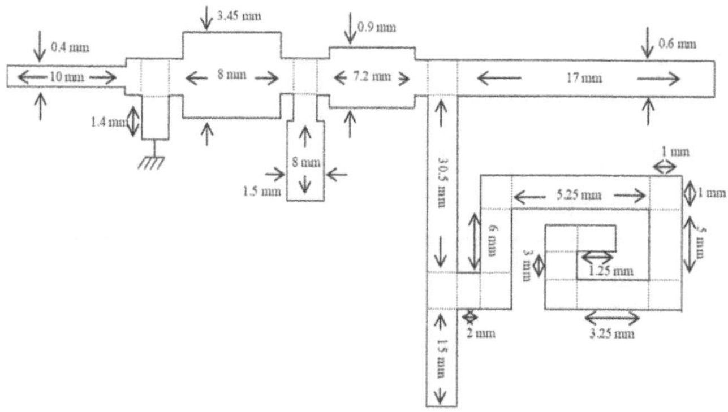

Figure 5.27. *Dimensions of the distributed TPMN using line sections*

Distributed TPMN

Owing to the reduced quality factor of the commercial lumped components and the unavailability of their exact required values, based on the lumped circuit shown in Figure 5.26(b), a distributed TPMN in microstrip technology is designed in order to maximize its performance. Duroid 5880 substrate with dielectric permittivity of 2.2, loss tangent of 0.0009 and thickness of 0.8 mm is used for the prototype with dimensions illustrated in Figure 5.27. The reflection coefficients of ports 1 and 3 are shown in Figure 5.28(a), and that of port 2 is shown in Figure 5.28(b). Transmission coefficients S31, S21 and S32 are shown in Figure 5.29. Their values are as expected, and they finally validate the operation principle of the TPMN.

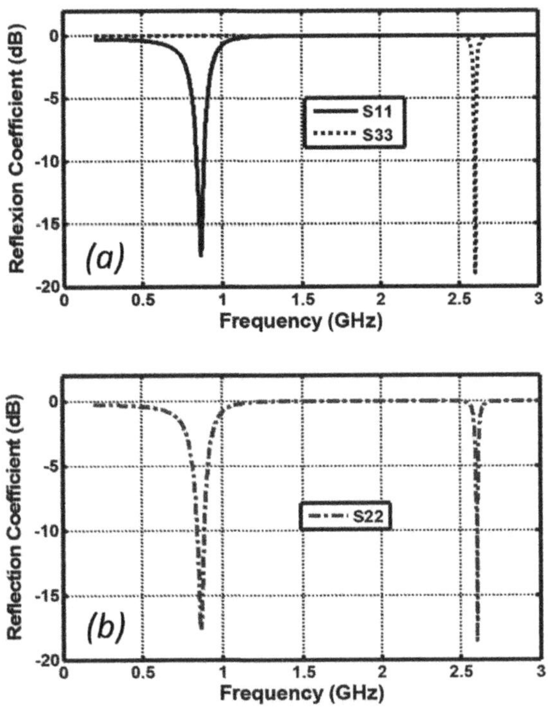

Figure 5.28. *Reflection coefficients at the inputs of the TPMN: a) S11 and S33, b) S22*

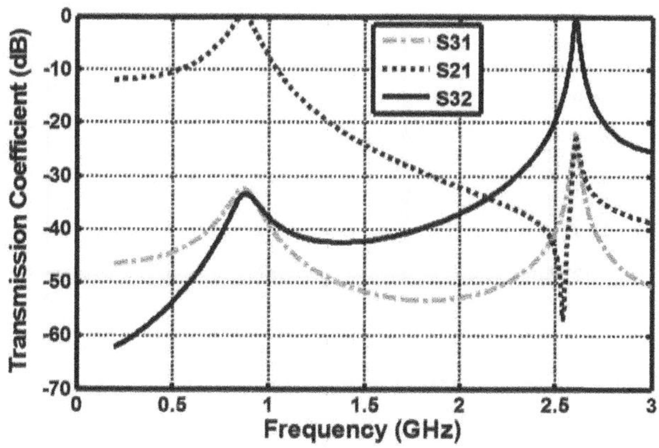

Figure 5.29. *Transmission coefficients of the TPMN: S21, S31 and S32*

5.3.3.2. *Harvesting the third harmonic*

The optimum harvesting of the third harmonic is performed by integrating the RFID section, HH section and the TPMN to operate all together. The integration is simulated using HB analysis. A parametric study of the operation of the augmented tag in a wide power range is performed by simulation. Setting a power sweep from –20 dBm to 10 dBm at the input at f_0 (i.e. port 1), the Power Spectral Density (PSD) at the input of the chip model (i.e. port 2) and at the input of the HH section (i.e. port 3) is reported in Figures 5.30 and 5.31, respectively.

Piled rows at a given frequency represent the power variations due to the sweep power setting. As an illustrative example, when the f_0 power injected in port 1 is 10 mW, the f_0 power conducted at the input of the chip model is 9.6 mW (Figure 5.30), which demonstrates a good TPMN operation transmitting from port 1 to port 2. With the same input power, the conducted power at the input of the HH section at $3f_0$ is 180 μW and the one coupled at f_0 is only 0.5 μW (Figure 5.31), which confirms the good TPMN isolation performance between ports 1 and 3 at f_0. The negative values shown in Figure 5.30 at $3f_0$ represent the direction of the signal with respect to the output of port 2 (negative values are entering into the TPMN). Its behavior is better observed in Figure 5.32, which shows the $3f_0$ power generated by the chip model and calculated at its input (i.e. port 2) as a function of the f_0 input power. When the f_0 input level is 10 mW, the $3f_0$ power reflected back into the TPMN and redirected towards the HH section (i.e. port 3) is 198 μW.

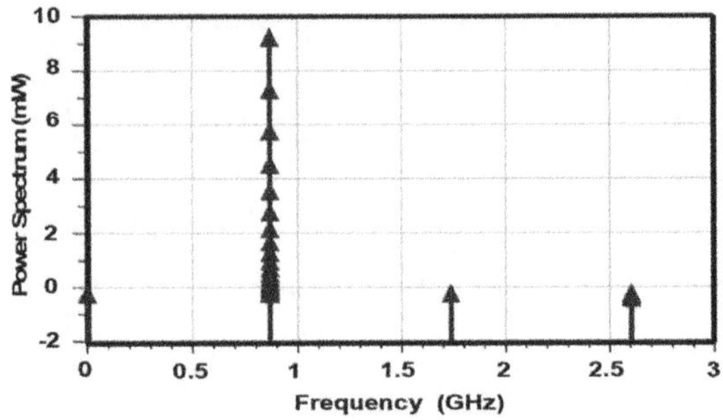

Figure 5.30. *Power spectral density at the input of the chip model (port 2)*

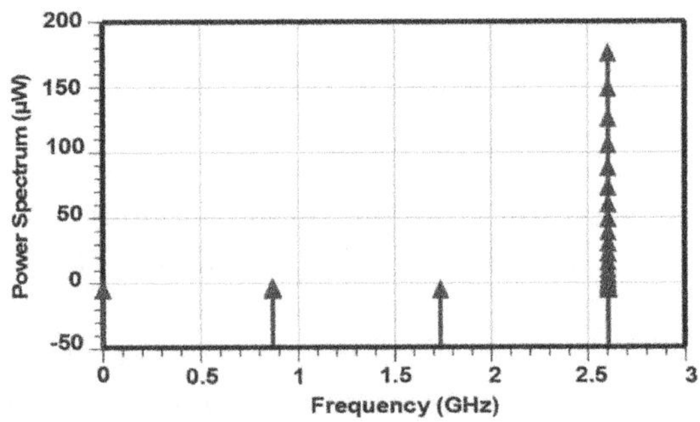

Figure 5.31. *Power spectral density at the input of the harvesting section (port 3)*

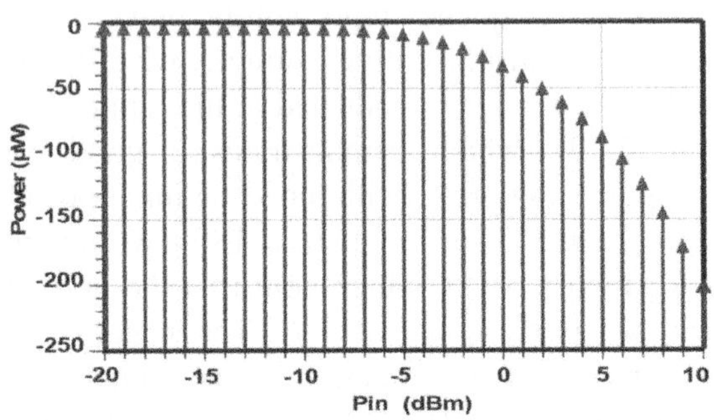

Figure 5.32. *The third harmonic RF power (µW) reflected at the entrance to the chip model versus the input power source (dBm)*

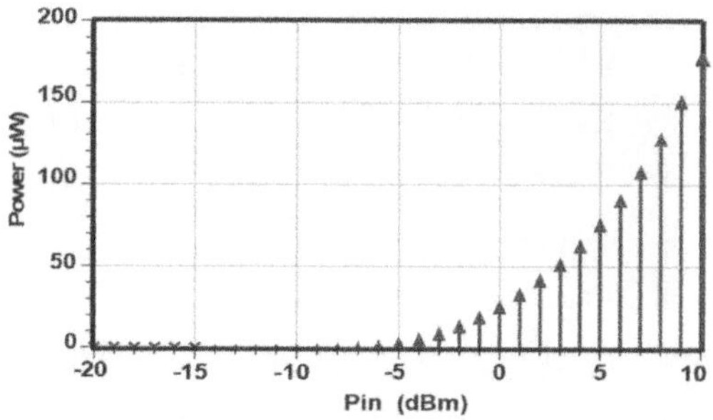

Figure 5.33. *The third harmonic RF power (μW) redirected at the entrance to the harvesting section versus the input power source (dBm)*

Figure 5.33, which unfolds the piled rows observed in Figure 5.31 at $3f_0$, represents the power variation entering into the HH section as a function of the f_0 input power. Positive values represent the signal flowing in the direction of the harvester. From the 198 μW reflected back at port 2, 180 μW enter into the HH section and 18 μW are lost in the transmission path (i.e. from port 2 to port 3). It is worth noting that these losses are non-linear and are significantly lower at low input power levels as shown in Figures 5.32 and 5.33. Once the operation of the TPMN, considering the RFID section (using the chip model) and the HH section connected, is validated, the RF-to-DC conversion efficiency of each section is optimized by analyzing the PSD at f_0 and at $3f_0$ at ports 2 and 3. The only optimization parameters are the TPMN dimensions and the value of C3. Optimized results are analyzed and compared with the experimental ones in the next section.

5.3.4. *Application case and experimental results*

5.3.4.1. *Evaluation of the chip model*

To experimentally validate the proposed design, the first step is to verify the behavior of the chip model compared with the real RFID chip sample. Two prototypes, one using the chip model and the other using the chip sample, are realized using a similar 50 Ω matching network at $f_0 = 0.868$ GHz as represented in Figure 5.34. Figure 5.35 compares the measured reflection coefficient (S11) of both prototypes. The results indicate a relatively good agreement despite a slight shift of the resonant frequency: at $f_0 = 0.868$ GHz, the S11 is –14 dB for the chip model and –11 dB for the real chip.

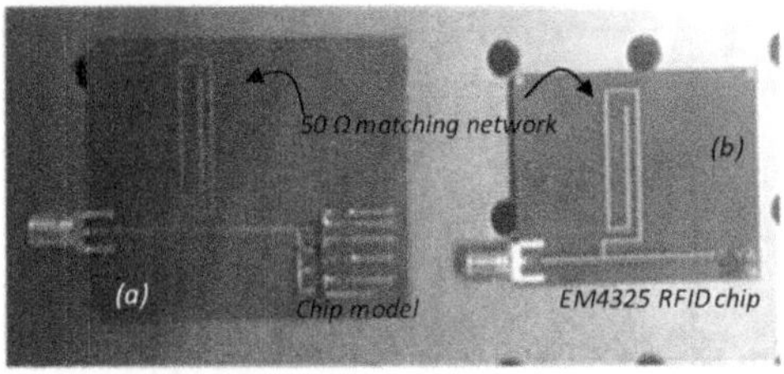

Figure 5.34. *Prototypes of (a) the RFID chip model and (b) the real RFID chip sample (EM4325) both matched to a 50 Ω input port*

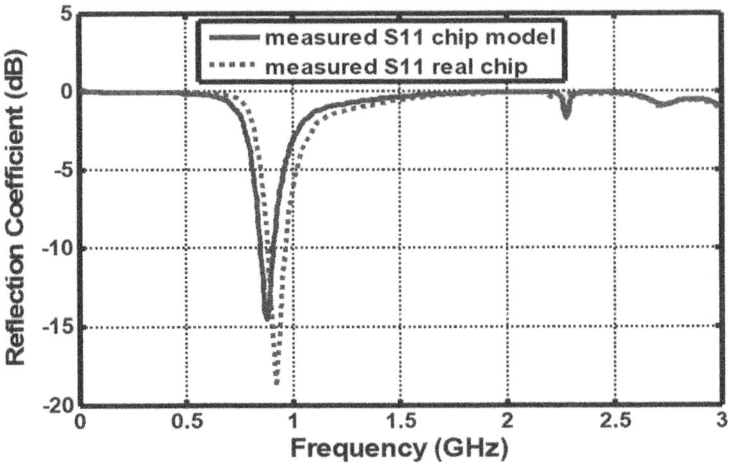

Figure 5.35. *Measured reflection coefficient S11 for chip model and real chip circuits matched to 50 Ω at 868 MHz*

5.3.4.2. *RF-to-DC conversion efficiency using the chip model*

The objective is to evaluate the performance of the HH section circuit in terms of RF-to-DC conversion efficiency using the equivalent model of the RFID chip. It is worth noting that the evaluation is only possible for the case with the chip model because the load of the commercial RFID chip is not accessible. For this study, the efficiency is evaluated by comparing the cases when only the RFID section is present using the chip model (i.e. the considered reference circuit) and when the RFID section using the chip model and the HH sections are present (i.e. augmented tag).

The conversion efficiency is classically defined as the ratio between the DC power (P_{dc}) consumed by the RL load and the RF power (P_{RF}) at the input of the circuit. For the reference circuit and the augmented tag, the respective efficiencies η_{ref} and η are distinguished by the corresponding P_{dc} as follows:

$$\eta_{ref} = \frac{P_{dc\_RL}}{P_{RF@0.868}} \quad [5.2]$$

$$\eta = \frac{P_{dc\_RL}+P_{dc\_RL3F}}{P_{RF@0.868}} \quad [5.3]$$

where $P_{RF@0.868}$ is the RF input power at 0.868 GHz, P_{dc_RL} is the power dissipated in the load of the chip model circuit and P_{dc_RL3F} is the power dissipated in the load of the $3f_0$ harvester. Figure 5.36 shows the two considered prototypes and the measurement setup composed of a Speedway RFID reader that generates a 0.868 GHz signal, an attenuator that is used to perform the analysis at an input power ranging from -20 dBm to 10 dBm, and a DC-meter voltage.

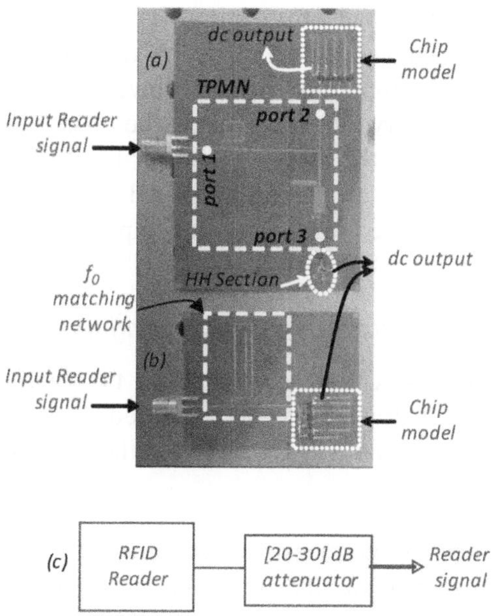

Figure 5.36. *Evaluation of the RF-to-DC conversion efficiency of the different prototypes: a) chip model; b) augmented tag that uses the chip model and c) measurement setup*

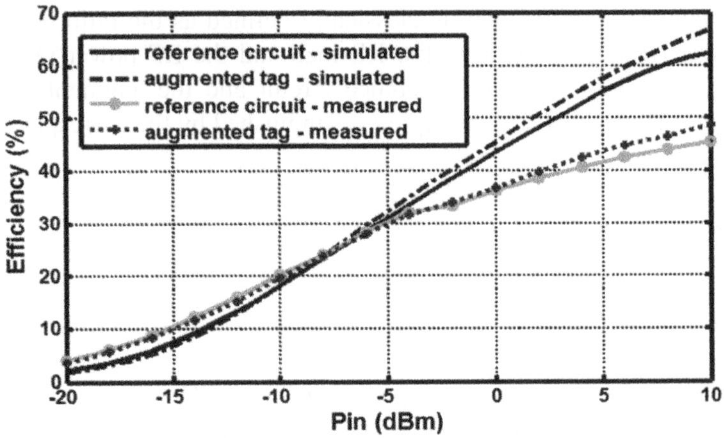

Figure 5.37. *RF-to-DC conversion efficiency of the RFID chip model compared with one of the augmented tags calculated as in (1) and (2), respectively*

Figure 5.37 shows the conversion efficiency as a function of the input power for the two considered prototypes. As the $3f_0$ level varies with the input power at f_0, the conversion efficiency of the reference circuit and that of the augmented tag for input power levels lower than –5 dBm is similar to each other. On the contrary, above –5 dBm, the efficiency is higher for the augmented tag because of the higher $3f_0$ power level produced by the chip. The maximum conversion efficiency measured for the reference circuit is 45%, and for the augmented tag (using the chip model), it is 48%, both at 10 dBm input power and 0.868 GHz.

5.3.4.3. *RF-to-DC conversion efficiency using a real RFID chip*

The objective of this section is to evaluate the harvesting performance of the $3f_0$ harvester. In this case, the chip model is replaced by the real chip sample using the same TPMN used for the model as it is shown in Figure 5.38.

Figure 5.39 compares the measured DC power harvested from $3f_0$ using the real RFID chip sample and the simulated one using the chip model, as a function of the f_0 input power at port 1.

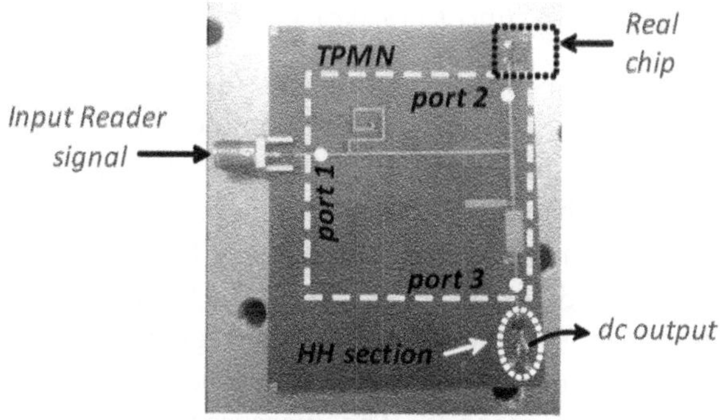

Figure 5.38. *Augmented tag using a real RFID chip*

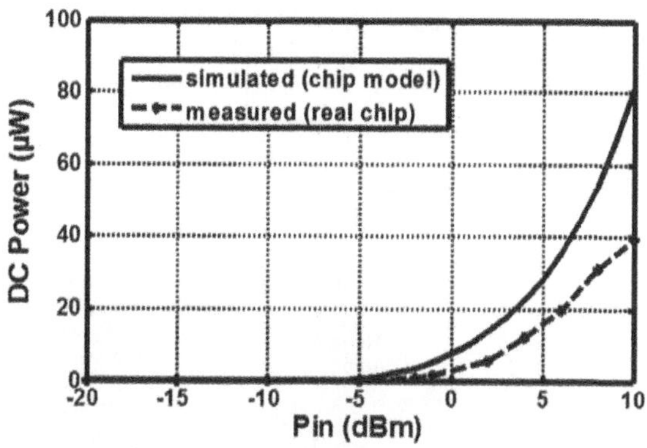

Figure 5.39. *DC power harvested from the third harmonic signal rectification as a function of the input power source (dBm)*

A maximum power of 39 μW is obtained with the augmented tag using the real chip, confirming the validity of the presented design methodology. The simulated conversion efficiency for the HH section can be calculated from Figure 5.33 (RF input power) and Figure 5.38 (DC output power); for instance, at 10 dBm input power and 0.868 GHz, the efficiency is 44%.

It is worth highlighting that the measured activation threshold of the augmented tag, using the setup of Figure 14, is –7.6 dBm, which represents an acceptable read range of 5.3 m as estimated in [RAO 05]. It is expected that, at this power level, the harvested energy from $3f_0$ will be almost null as shown in Figure 5.33.

5.3.4.4. *Harmonic harvesting to power a temperature sensor*

Finally, the objective is now to exploit the sole harmonic harvested power by the augmented tag to energize a commercial temperature sensor (PSG03551), besides establishing a standard RFID communication. The operation is possible thanks to the use of a power management module (BQ25504EVM-674) that performs three main functions: (i) the DC–DC boosting (active above 320 mV at the input); (ii) the power storage (using a 100 μF capacitor); and (iii) the output voltage regulation (set to deliver 1 V at the output). The temperature sensor consumes a power of 60 μW in sensing mode every 10 seconds. The measured input impedance of the power management module connected in series with the sensor is 3.3 kΩ.

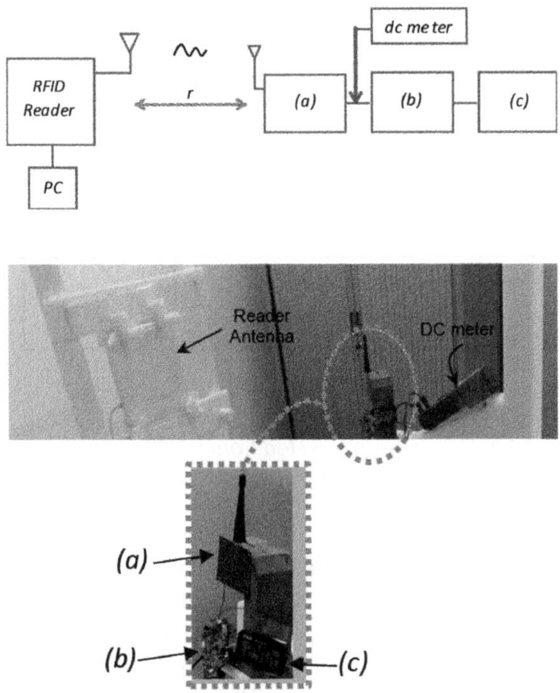

Figure 5.40. *Evaluation setup of the augmented tag supplying a commercial temperature sensor: a) augmented tag; b) power management module and c) temperature sensor*

The setup used in order to evaluate the sensing application in harsh conditions is depicted in Figure 5.39. A Speedway RFID reader driven by a portable computer (PC) is used to communicate with the augmented tag. Regarding the tag side configuration, a 50 Ω UHF monopole antenna is connected to the port 1 of the prototype using the real RFID chip, the DC output of the HH section is connected to the DC input of the power management module, and its output is connected to the temperature sensor. In addition, a DC meter is used to measure the voltage delivered by the HH section. With a distance of 80 cm between the antenna reader and the prototype, and setting an EIRP of 35 dBm in the reader, the HH section is able to deliver 320 mV of DC voltage and the power management module starts operating and activates the sensor. It is worth noting that at distances higher than 80 cm, the sensor is not activated but the RFID communication remains established up to a distance of 5 m.

5.3.5. *Summary: harmonic harvesting*

Section 5.3 has presented a relevant new concept based on the energy harvesting of wasted harmonic signals produced by standard passive RFID UHF chips. The approach involves the exploitation of the third harmonic signal generated by a traditional passive RFID chip to produce an additional source of power. To design the proposed system, a diode-based voltage multiplier circuit is used as the non-linear RF front-end model of a passive UHF RFID chip. Furthermore, a specific three-port matching network is designed and optimized to properly distribute the different involved signals, i.e. to obtain an RFID tag able to provide DC power to an associated sensor at the same time as communicating with the reader. The presented simulations allow the detailed study of the device integration, in particular the efficient redirection of the third harmonic signal from the RFID chip to the harmonic harvester circuit. The introduced models for the chip and the complete system are also experimentally validated. The system performance is fully characterized in terms of DC power and RF-to-DC conversion efficiency. Simulated results agree well with the measurement.

In terms of performance, considering a 3.3 kΩ load, a DC power of 39 μW is harvested from the third harmonic produced by the RFID chip, at 10 dBm input power at the fundamental frequency of 868 MHz in Europe. Finally, the proposed system is used to supply DC power to a low consumption temperature sensor. All this was performed without any other energy source than the third harmonic produced by the RFID chip.

Moreover, the concept demonstrated here can be adapted to any other RFID chip once it is accurately characterized. As the sensitivity of recent chips can be better than –22 dBm [IMP 16, IMP 17], it is expected that different scenarios of application can be developed, including the harvesting of the weak third harmonic emitted by the reader. Besides, more complexity and compactness can be introduced in the RF matching section. Among the possible applications, the harmonic harvesting can be used for passive beam steering in the tag side. Indeed, as explained in [NIK 17], by switching the load of a parasitic radiative element between capacitive and inductive impedances, the beam of a director element, tuned next to it, can be steered in two opposite directions, i.e. the parasitic switches between the director and the reflector. The passive beam steering in the tag could solve many reading problems when dense tag sceneries prevent 100% tag detection. In this context, a low-consumption switching circuit like the one in [NIK 17] can be feed harvesting the harmonic RFID signals.

5.4. Conclusion

This chapter discussed the exploitation of non-linear RFID chip characteristics to design some augmented tags that are able to achieve more functions than standard UHF tags. Of course, the developed augmented tags are fully compliant with existing regulations and standards, in particular the EPC C1 GEN2 [EPC 15]. To design augmented tags, we exploited the third harmonic "naturally" generated by the rectifier section of the RFID chip. Two quite different study cases have been considered: harmonic tag and harvesting tag.

The new functionality added by the HT is primarily channel diversity. Indeed, without any change in the architecture of the RFID chip, and adopting a proper design of the tag antenna to make it operate at both fundamental and third harmonic frequency, the tag will backscatter its information at the third harmonic too. Therefore, a channel diversity is possible with HT. Several scenarios exploiting this channel diversity can be defined. They are primarily enabled by the specific design of the dual band tag antenna. It is remarkable to note that the read-range at the harmonic channel is quite comparable to the read range at the fundamental frequency, as it has demonstrated by measurement.

The new functionality added by the harvesting tag regards the collection of a certain amount of power that is usually wasted. Such "additional" power is mainly the power generated at the third harmonic frequency due to the non-linear response of the rectifier of the RFID chip. At least three different scenarios can be envisaged. In the first scenario, the power at the third harmonic is rectified and re-injected to empower the RFID tag, for instance, to the DC input of chips that have a battery-assisted port. In the second scenario, the generated third harmonic is first mixed with an external signal like Wi-Fi before rectification and finally feeding the battery-assisted port. Such a mixing operation allows the optimization of the conversion-to-DC efficiency as it has been demonstrated in [COL 14]. In the third scenario, the generated harmonic signal is collected via a passive matching network, then rectified and the DC power is used to feed an external circuit like a low power sensor. This concept has been demonstrated with a commercial temperature sensor.

It is worth noting that the field of application is larger than the examples discussed herein. While the demonstrations presented in this chapter are based on a standard RFID chip, it is obvious that RFID chips can be designed in order to optimize the exploitation of these harmonics. Despite the constraints and technology limitations, the future looks very promising for the RFID technology because of its ability to implement the IoT.

Conclusion

An Old Technology with a Bright Future

As Harry Stockman concluded in his landmark paper published in 1948, "Evidently, considerable research and development work has to be done before the remaining basic problems in reflected-power communication are solved, and before the field of useful applications is explored." Nowadays, even if RFID is a widespread technology, it also has solid arguments and remarkable potential in terms of new applications and unexpected new functionalities. More than seven decades after the invention of the "Thing" by L. Theremin, the evolution of RFID, from an identification technique to a more ubiquitous and context-aware technology, has been growing very rapidly, in particular since the 2000s.

After the early RFID devices developed in the 1960s for electronic article surveillance (EAS), implanted tags for tattooing and identification of pets, or toll systems, several technology advances allowed the integration of all the RFID circuitry in the passive chip. One of the major developments was that an UHF RFID system was performed and patented in the 1990s by IBM engineers. The same decade witnessed the emergence of standards, which favored the deployment of RFID, in particular for logistics and traceability sectors. Some years later, RFID became a part of our everyday life with applications in ticketing, access control and e-papers, and so on.. In all these developments and applications, only the capabilities of unitary identification of RFID were exploited.

Coupling the RFID with other technologies such as Wi-Fi, Bluetooth and GPS decoupled the potential of RFID applications, known before as a simple identify enabler, and initiated the pre-birth of a ubiquitous and context-aware system. Indeed, data fusion and processing allowed the engineers to build smart applications which are able to provide more information as well as to deliver some predictions that help to make decisions and take specific actions. In other words, it is neither more nor less than the concept of the Internet of Things (IoT).

The implementation of the IoT in the real world requires the availability of a huge number of node devices, probably some tens of billions. Such nodes, in addition to unitary identification capability, should exhibit important features such as wireless communication, no battery, sensing capabilities and localization, potentially based on green technologies and recyclability. Hopefully, they should be as cost-effective as soft and flexible labels. Considering the previous features, the only technology that can fit them nowadays is the passive RFID technology, in particular in the UHF band, for which the read range has recently exceeded 28 m, enabling plenty of applications in indoor environments. So, UHF RFID is probably the cornerstone for the implementation of the last few meters of the IoT.

To acquire new functionalities, the hardware and software components of the RFID tag should be reinforced and passive ID tags transformed into augmented tags. Three complementary methods can be considered for such evolution.

The first method consists of adding specific hardware and circuitry which are able to provide extra functions such as sensing. In most cases, such an approach requires additional power to feed the added circuitry; therefore, the important feature of a batteryless solution is no longer possible.

The second method exploits the properties of the tag antenna and its sensitivity to any change in its near-field region. Such a change can be reinforced by loading the antenna with a material sensitive to environmental parameters such as moisture, temperature and pressure. This capability has been demonstrated by several research groups worldwide. Even if this approach is more complex and more challenging than the first method, it continues to be considered in several R&D projects.

The third method is based on exploiting the characteristics of the RFID chip, in particular its non-linear response that is generated by the rectifier section of the chip. This is the topic discussed in this book. It is demonstrated that it is necessary to consider the generation of harmonic signals by the rectifier with sufficient power for enabling new functions of the RFID tag, without losing any of the standard duties of the tag. By exploiting the third harmonic signal, this book reports new functions such as channel diversity, energy harvesting and sensor feeding. This third tag transformation method appears to be more profitable and rich in terms of applications. Indeed, the recent advances in RFID chip technology, especially in terms of better sensitivity and new functions such as auto-tuning capability, will enable the design of augmented tags and allow the implementation of unprecedented smart applications in the near future. This is why we can claim that RFID is "an old technology with a bright future".

Bibliography

[AAZ 09] AAZOU S., ASSAID E.-M., "Schottky diode parameters extraction using two different methods", *2009 International Conference on Microelectronics (ICM)*, pp. 240–243, December 2009.

[AGI 06] AGILENT TECH NOLOGIES, Agilent electronic calibration (ECal) modules for vector network analyzers, available at: http://www3.imperial.ac.uk/pls/portallive/docs/1/11953697.pdf, 2006.

[AGI 11] AGILENT TECHNOLOGIES, Harmonic balance simulation, available at: http://cp.literature.agilent.com/litweb/pdf/ads2006/pdf/cktsimhb.pdf, 2011.

[ALI 10] ALIMENTI F., PALAZZARI V., ORECCHINI G. *et al.*, "Crossed dipole frequency doubling RFID TAG based on paper substrate and ink-jet printing technology", *Microwave Symposium Digest (MTT), 2010 IEEE MTT-S International*, pp. 840–842, May 2010.

[AND 13a] ANDIA VERA G., DUROC Y., TEDJINI S., "Analysis of harmonics in UHF RFID signals", *IEEE Transactions on Microwave Theory and Techniques*, vol. 61, no. 6, pp. 2481–2490, 2013.

[AND 13b] ANDIA VERA G., DUROC Y., TEDJINI S., "Redundant backscattering modulation of passive UHF RFID tags", *Microwave Symposium Digest (MTT), 2013 IEEE MTT-S International*, pp. 1–3, June 2013.

[AND 14a] ANDIA VERA G., DUROC Y., TEDJINI S., "RFID test platform: non-linear characterization", *IEEE Transactions on Instrumentation and Measurement*, vol. 63, no. 9, pp. 2299–2305, doi: 10.1109/TIM.2014.2307754, September 2014.

[AND 14b] ANDIA VERA G., DUROC Y., TEDJINI S., "Tag-to-reader harmonic link in passive UHF RFID", *2014 IEEE MTT-S International Microwave Symposium Digest (MTT)*, pp. 1–3, June 2014.

[AND 17] ANDIA G., "Slenderly and conformable passive UHF RFID yarn", *IEEE International Conference on RFID (RFID)*, pp. 130–136, Phoenix, doi: 10.1109/RFID.2017.7945598, 2017.

[ATH 13] ATHALYE A., SAVIC V., BOLIC M. *et al.*, "Novel semi-passive RFID system for indoor localization", *IEEE Sensors Journal*, vol. 13, no. 2, pp. 528–537, 2013.

[AVA 09] AVAGO TECHNOLOGIES, Surface Mount Microwave Schottky Detector Diodes – Datasheet, available at: https://docs.broadcom.com/docs/AV02-1388EN, 2009.

[AVE 09] AVERY DENNISON RFID, AD-222 RFID Inlays-Datasheet, available at: http://www.sysscan.com.hk/PDF/AveryDennison_AD-222.pdf, 2009.

[AVO 05] AVOINE G., OECHSLIN P., "RFID traceability: a multilayer problem", *Proceedings of the 9th International Conference on Financial Cryptography and Data Security*, pp. 125–140, Springer-Verlag, 2005.

[BAB 14] BABOLI E.Y.M., SINGH A., BORIC-LUBECKE O. *et al.*, "E-healthcare: remote monitoring, privacy, and security", *2014 IEEE MTT-S International Microwave Symposium Digest (MTT)*, pp. 1–3, June 2014.

[BAL 05] BALANIS C.A., *Antenna Theory: Analysis and Design*, Wiley-Interscience, 2005.

[BAR 06] BARNETT R.E., LAZAR S., LIU J., "Design of multistage rectifiers with low-cost impedance matching for passive RFID tags", *2006 IEEE Radio Frequency Integrated Circuits (RFIC) Symposium*, p. 4, June 2006.

[BAR 09] BARNETT R.E., LIU J., LAZAR S., "A RF to dc voltage conversion model for multi-stage rectifiers in UHF RFID transponders", *IEEE Journal of Solid-State Circuits*, vol. 44, no. 2, pp. 354–370, 2009.

[BOA 11] BOAVENTURA A.S., CARVALHO N.B., "Maximizing dc power in energy harvesting circuits using multisine excitation", *Microwave Symposium Digest (MTT), 2011 IEEE MTT-S International*, pp. 1–4, June 2011.

[BOA 13] BOAVENTURA A., COLLADO A., CARVALHO N.B. *et al.*, "Optimum behavior: wireless power transmission system design through behavioral models and efficient synthesis techniques", *IEEE Microwave Magazine*, vol. 14, no. 2, pp. 26–35, 2013.

[BOL 10] BOLOMEY J.C., CAPDEVILA S., JOFRE L. *et al.*, "Electromagnetic modeling of RFID-modulated scattering mechanism. Application to tag performance evaluation", *Proceedings of the IEEE*, vol. 98, no. 9, pp. 1555–1569, September 2010.

[BOR 14] BORGES CARVALHO N., GEORGIADIS A., COSTANZO A. *et al.*, "Wireless Power transmission: R&D activities within Europe", *IEEE Transactions on Microwave Theory and Techniques*, vol. 62, no. 4, pp. 1031–1045, 2014.

[BRA 30] BRARD E., Patent US1744036A, 1930.

[BRO 00] BROWN L., *A Radar History of World War II: Technical and Military Imperatives*, Institute of Physics Publishing, 2000.

[CAB 04] CABRIA L., GARCIA J.A., MALAVER E. *et al.*, "A PHEMT frequency doubling active antenna with BPSK modulation capability", *IEEE Antennas and Wireless Propagation Letters*, vol. 3, no. 1, pp. 310–313, 2004.

[CAP 15] CAPDEVILA CASCANTE S. *et al.*, RFID multiantenna systems for wireless communications and sensing, Polytechnic University of Catalonia, available at: http://upcommons.upc.edu/handle/2117/94892, 2015.

[CAT 11] CATARINUCCI L., DE DONNO D., GUADALUPI M. *et al.*, "Performance analysis of passive UHF RFID tags with GNU-radio", *2011 IEEE International Symposium on Antennas and Propagation (APSURSI)*, pp. 541–544, 2011.

[CAT 12] CATARINUCCI L., DE DONNO D., COLELLA R. *et al.*, "A cost-effective SDR platform for performance characterization of RFID tags", *IEEE Transactions on Instrumentation and Measurement*, vol. 61, no. 4, pp. 903–911, 2012.

[CHA 89] CHAMBERLIN K., "Quantitative analysis of intermodulation product interference", *IEEE Transactions on Electromagnetic Compatibility*, vol. 31, no. 3, pp. 316–319, 1989.

[CHE 00] CHENG Y.T., LIN L., NAJAFI K., "Localized silicon fusion and eutectic bonding for MEMS fabrication and packaging", *Journal of Microelectromechanical Systems*, vol. 9, no. 1, pp. 3–8, doi: 10.1109/84.825770, March 2000.

[CHE 09] CHEN S.-L., LIN K.-H., "Characterization of RFID strap using single-ended probe", *IEEE Transactions on Instrumentation and Measurement*, vol. 58, no. 10, pp. 3619–3626, 2009.

[CHE 10] CHEN S.-L., LIN K.-H., MITTRA R., "A measurement technique for verifying the match condition of assembled RFID tags", *IEEE Transactions on Instrumentation and Measurement*, vol. 59, no. 8, pp. 2123–2133, 2010.

[COL 98] COLBURN J.S., JENSEN M.A., RAHMAT-SAMII Y., "Indoor ISM band multipath fading: frequency and antenna diversity", *1998 IEEE-APS Conference on Antennas and Propagation for Wireless Communications*, pp. 9–12, November 1998.

[COL 04] COLPITTS B.G., BOITEAU G., "Harmonic radar transceiver design: miniature tags for insect tracking", *IEEE Transactions on Antennas and Propagation*, vol. 52, no. 11, pp. 2825–2832, 2004.

[COL 12] COLLADO A., GEORGIADIS A., "Improving wireless power transmission efficiency using chaotic waveforms", *2012 IEEE MTT-S International Microwave Symposium Digest (MTT)*, pp. 1–3, June 2012.

[COL 14] COLLADO A., GEORGIADIS A., "Optimal waveforms for efficient wireless power transmission", *IEEE Microwave and Wireless Components Letters*, vol. 24, no. 5, pp. 354–356, 2014.

[COO 14] COOK B.S., VYAS R., KIM S. *et al.*, "RFID-based sensors for zero-power autonomous wireless sensor networks", *IEEE Sensors Journal*, vol. 14, no. 8, pp. 2419–2431, 2014.

[CUR 05] CURTY J.-P., JOEHL N., DEHOLLAIN C. *et al.*, "Remotely powered addressable UHF RFID integrated system", *IEEE Journal of Solid-State Circuits*, vol. 40, no. 11, pp. 2193–2202, 2005.

[DED 13] DE DONNO D., RICCIATO F., TARRICONE L., "Listening to tags: uplink RFID measurements with an open-source software-defined radio tool", *IEEE Transactions on Instrumentation and Measurement*, vol. 62, no. 1, pp. 109–118, 2013.

[DED 14] DE DONNO D., CATARINUCCI L., TARRICONE L., "A battery-assisted sensor-enhanced RFID tag enabling heterogeneous wireless sensor networks", *IEEE Sensors Journal*, vol. 14, no. 4, pp. 1048–1055, April 2014.

[DEE 10] DEEPU V., VENA A., PERRET E. *et al.*, "New RF identification technology for secure applications", *2010 IEEE International Conference on RFID-Technology and Applications (RFID-TA)*, pp. 159–163, June 2010.

[DEV 05a] DE VITA G., IANNACCONE G., "Design criteria for the RF section of UHF and microwave passive RFID transponders", *IEEE Transactions on Microwave Theory and Techniques*, vol. 53, no. 9, pp. 2978–2990, 2005.

[DEV 05b] DE VITA G., IANNACCONE G., "Ultra low power RF section of a passive microwave RFID transponder in 0.35 mu;m BiCMOS", *IEEE International Symposium on Circuits and Systems ISCAS 2005*, vol. 5, pp. 5075–5078, May 2005.

[DIC 76] DICKSON J.F., "On-chip high-voltage generation in MNOS integrated circuits using an improved voltage multiplier technique", *IEEE Journal of Solid-State Circuits*, vol. 11, no. 3, pp. 374–378, 1976.

[DIL 15] DI LONARDO R., "Electronic article surveillance (EAS) source tagging: 20+ years of innovation", available at: http://losspreventionmedia.com/insider/loss-prevention-technology/electronic-article-surveillance-eas-source-tagging/, 2015.

[DOB 05] DOBKIN D.M., WEIGAND S.M., "Environmental effects on RFID tag antennas", *2005 IEEE MTT-S International Microwave Symposium Digest*, p. 4, June 2005.

[DOB 08] DOBKIN D.M., *The RF in RFID. Passive UHF RFID in Practice*, Elsevier, 2008.

[DUR 12] DUROC Y., KADDOUR D., "RFID potential impacts and future evolution for green projects", *Energy Procedia*, vol. 18, pp. 91–98, 2012.

[DUR 13] DUROC Y., ANDIA VERA G., GARCIA MARTIN J.P., "Modified RSSI technique for the localization of passive UHF RFID tags in LOS channels", *International Journal of Microwave and Wireless Technologies*, vol. 5, no. 5, pp. 645–651, 2013.

[DUR 14] DUROC Y., ANDIA VERA G., "Towards autonomous wireless sensors: RFID and energy harvesting solutions", in MUKHOPADHYAY S.C., *Internet of Things*, Springer, 2014.

[DUR 18] DUROC Y., TEDJINI S., *RFID a Key Technology for Humanity*, CRAS, 2018.

[ECC 12] ECC, Compatibility between the mobile service in the band 2500–2690 MHz and the radiodetermination service in the band 2700–2900 MHz, Report no. 174, available at: http://www.erodocdb.dk/docs/doc98/official/pdf/ECCRep174.pdf, 2012.

[EMM 13] EM-MICROELECTRONIC-MARIN-SA, EM4325 datasheet, available at: http://www.emmicroelectronic.com/sites/default/files/public/products/datasheets/em4325\s\do3(d)s.pdf, 2013.

[EPC 15] EPC GLOBAL STANDARDS, EPC UHF Gen2 Air Interface Protocol, available at: https://www.gs1.org/epcrfid/epc-rfid-uhf-air-interface-protocol/2-0-1, 2015.

[ETS 08] ETSI, Electromagnetic compatibility and Radio spectrum Matters (ERM); Radio Frequency Identification Equipment operating in the band 865 MHz to 868 Hz with power levels up to 2 W; Part 1: Technical requirements and methods of measurement, EN 302 208-1 V1.2.1, available at: http://www.etsi.org/deliver/etsi_en/302200_302299/30220801/01.02.01_60/en_30220801v010201p.pdf, 2008.

[FAL 12] FALKENSTEIN E., ROBERG M., POPOVIC Z., "Low-power wireless power delivery", *IEEE Transactions on Microwave Theory and Techniques*, vol. 60, no. 7, pp. 2277–2286, 2012.

[FED 15] FEDERAL COMMUNICATIONS COMMISSION, Guidelines for determining the effective radiated power (ERP) and equivalent isotropically radiated power (EIRP) of a RF transmitting system, available at: https://apps.fcc.gov/oetcf/kdb/forms/FTSSearchResultPage.cfm?switch=P&id=47469, 2015.

[FON 82] FONG E., ZEMAN R., "Analysis of harmonic distortion in single-channel MOS integrated circuits", *IEEE Journal of Solid-State Circuits*, vol. 17, no. 1, pp. 83–86, 1982.

[FOS 07] FOSTER K.R., JAEGER J., "RFID Inside", *IEEE Spectrum*, vol. 44, no. 3, pp. 24–29, doi: 10.1109/MSPEC.2007.323430, March 2007.

[GEO 10] GEORGIADIS A., ANDIA G., COLLADO A., "Rectenna design and optimization using reciprocity theory and harmonic balance analysis for electromagnetic (EM) energy harvesting", *IEEE Antennas and Wireless Propagation Letters*, vol. 9, pp. 444–446, 2010.

[GEO 12] GEORGIADIS A., COLLADO A., "Improving range of passive RFID tags utilizing energy harvesting and high efficiency class-E oscillators", *2012 6th European Conference on Antennas and Propagation (EUCAP)*, pp. 3455–3458, March 2012.

[GOM 07] GOMES H.C., CARVALHO N.B., "The use of intermodulation distortion for the design of passive RFID", *European Radar Conference, 2007. EuRAD 2007*, pp. 377–380, October 2007.

[GOM 09] GOMES H.C., CARVALHO N.B., "RFID for location proposes based on the intermodulation distortion", *Sensors and Transducers Journal*, vol. 106, no. 7, pp. 85–96, 2009.

[GOR 14] GORDONSON J., BARTON T.W., PERREAULT D.J., "Transmission line resistance compression networks for microwave rectifiers", *Microwave Symposium Digest (MTT), 2014 IEEE MTT-S International*, pp. 1–3, June 2014.

[HAR 60] HARRIS D.B., Radio transmission systems with modulatable passive responder, US Patent 2,927,321, March 1, 1960.

[HAR 92] HARRISON R.G., "Full non-linear analysis of detector circuits using Ritz-Galerkin theory", *IEEE MTT-S International Microwave Symposium Digest, 1992*, vol. 1, pp. 267–270, June 1992.

[HEM 14] HEMOUR S., WU K., "Radio-frequency rectifier for electromagnetic energy harvesting: development path and future outlook", *Proceedings of the IEEE*, vol. 102, no. 11, pp. 1667–1691, 2014.

[HIN 84] HINES M.E., "The virtues of non-linearity–detection, frequency conversion, parametric amplification and harmonic generation", *IEEE Transactions on Microwave Theory and Techniques*, vol. 32, no. 9, pp. 1097–1104, 1984.

[HIR 06] HIRVONEN M., JAAKKOLA K., PURSULA P. *et al.*, "Dual-band platform tolerant antennas for radio-frequency identification", *IEEE Transactions on Antennas and Propagation*, vol. 54, no. 9, pp. 2632–2637, 2006.

[HUE 06] HUELSMEYER C., Huelsmeyer's "telemobiloscope", US patent 810,150, January 16, 1906.

[IMP 16] IMPINJ, Monza 4 Tag Chip Datasheet, available at: http://www.rfidconnect.com/mediaservice.ashx?type=attachmentid=4cff4c28-355c-405d-9afd-ad714b8e42d4, 2016.

[IMP 17] IMPINJ, Impinj UHF RFID Product downloads - Monza tag chips, available at: http://www.impinj.com, 2017.

[ISO 12] ISO/IEC 18000-1, Information technology AIDC techniques RFID for item management air interface, part 1 generic parameters for air interface communication for globally accepted frequencies, available at: http://www.iso.org, 2012.

[ISO 15] ISO/IEC 18000, Automatic identification radio frequency identification for item management communications and interfaces, available at: http://www.iso.org, 2015.

[JAL 05] JALALY I., ROBERTSON I.D., "Capacitively-tuned split microstrip resonators for RFID barcodes", *2005 European Microwave Conference*, vol. 2, p. 4, October 2005.

[JAN 88] JANASWAMY R., LEE S.-W., "Scattering from dipoles loaded with diodes", *IEEE Transactions on Antennas and Propagation*, vol. 36, no. 11, pp. 1649–1651, 1988.

[JEO 06] JEON S., YU Y., CHOI J., "Dual-band slot-coupled dipole antenna for 900MHz and 2.45 GHz RFID tag application", *Electronics Letters*, vol. 42, no. 22, pp. 1259–1260, 2006.

[KAR 03] KARTHAUS U., FISCHER M., "Fully integrated passive UHF RFID transponder IC with 16.7-mu; W minimum RF input power", *IEEE Journal of Solid-State Circuits*, vol. 38, no. 10, pp. 1602–1608, 2003.

[KEY 14] KEYSIGHT TECHNOLOGIES, Triggering PNA microwave network analyzers for antenna measurements, White Paper, available at: http://cp.literature.agilent.com/litweb/pdf/5988-9518EN.PDF, 2014.

[KLA 03] FINKENZELLER K., *RFID Handbook: Fundamentals and Applications in Contactless Smart Cards and Identification*, 2nd ed., John Wiley & Sons, New York, 2003.

[LAD 13] LADAN S., GHASSEMI N., GHIOTTO A. *et al.*, "Highly efficient compact rectenna for wireless energy harvesting application", *IEEE Microwave Magazine*, vol. 14, no. 1, pp. 117–122, 2013.

[LAD 14] LADAN S., WU K., "35 GHz harmonic harvesting rectifier for wireless power Transmission", *2014 IEEE MTT-S International Microwave Symposium (IMS)*, pp. 1–4, June 2014.

[LAN 05] LANDT J., "The history of RFID", *IEEE Potentials*, vol. 24, no. 4, pp. 8–11, October–November 2005.

[LAS 09] LASSER G., LANGWIESER R., SCHOLTZ A.L., "Broadband suppression properties of active leaking carrier cancellers", *2009 IEEE International Conference on RFID*, pp. 208–212, April 2009.

[LCI 17] LCIS, "Laboratoire de Conception et dIntegration des Systemes", available at: http://lcis.grenoble-inp.fr/le-laboratoire, 2017.

[LEE 11] LEE M.-H., YAO C.-Y., LIU H.-C., "Passive Tag for Multi-carrier RFID Systems", *2011 IEEE 17th International Conference on Parallel and Distributed Systems (ICPADS)*, pp. 872–876, December 2011.

[LIU 07a] LIU H.-C., CHEN Y.-F., CHEN Y.-T., "A frequency diverse Gen2 RFID system with isolated continuous wave emitters", *Science*, vol. 2, no. 5, pp. 54–60, 2007.

[LIU 07b] LIU H.-C., GUO X.-C., TZENG W.-S., "A UHF passive RFID system with frequency diversity", *2007 International Symposium on Antennas and Propagation (ISAP2007)*, pp. 20–24, 2007.

[LIU 07c] LIU H.-C., CHEN Y.-T., TZENG W.-S., "A multi-carrier UHF passive RFID system", *SAINT Workshops 2007. International Symposium on Applications and the Internet Workshops, 2007*, IEEE, pp. 21–21, 2007.

[LIU 09] LIU H.-C., CIOU J.-P., "Performance analysis of multi-carrier RFID systems", *SPECTS 2009. International Symposium on Performance Evaluation of Computer & Telecommunication Systems, 2009*, IEEE, vol. 41, pp. 112–116, 2009.

[LIU 10] LIU J., "Dual-band RFID tag antenna using coplanar inverted-F/L structure", *2010 IEEE International Conference on RFID-Technology and Applications (RFID-TA)*, pp. 96–99, June 2010.

[LIU 11] LIU H.-C., LIN W.-C., LIN M.-Y. *et al.*, "Passive UHF RFID tag with backscatter diversity", *Antennas and Wireless Propagation Letters, IEEE*, vol. 10, pp. 415–418, 2011.

[LON 12] MA Z.L., JIANG L.J., XI J. *et al.*, "A single-layer compact HF-UHF dual-band RFID tag antenna", *IEEE Antennas and Wireless Propagation Letters*, vol. 11, pp. 1257–1260, 2012.

[LUK 07] MAYER L.W., SCHOLTZ A.L., "Sensitivity and impedance measurements on UHF RFID transponder chips", *International EURASIP Workshop on RFID Technology*, pp. 1–10, 2007.

[MA 14] MA Y., KAN E.C., "Accurate indoor ranging by broadband harmonic generation in passive NLTL Backscatter tags", *IEEE Transactions on Microwave Theory and Techniques*, vol. 62, no. 5, pp. 1249–1261, 2014.

[MAA 03] MAAS S.A., *Non-linear Microwave and RF Circuits*, Artech House, 2003.

[MAL 11] MALHEIROS-SILVEIRA G.N., MORETTI A., HERNANDEZ-FIGUEROA H.E. *et al.*, "Exploring dual-band RFID tag antennas by means of asymmetric dipoles", *Microwave Optoelectronics Conference (IMOC), 2011 SBMO/IEEE MTT-S International*, pp. 244–248, October 2011.

[MAR 08] MARROCCO G., "The art of UHF RFID antenna design: impedance-matching and size-reduction techniques", *IEEE Antennas and Propagation Magazine*, vol. 50, no. 1, pp. 66–79, 2008.

[MAR 10] MARROCCO G., "Pervasive electromagnetics: sensing paradigms by passive RFID technology", *IEEE Wireless Communication*, vol. 17, no. 6, pp. 10–17, 2010.

[MAR 14] MARIOTTI C., ORECCHINI G., ALIMENTI F. *et al.*, "7.5–15 MHz organic frequency doubler made with pentacene-based diode and paper substrate", *2014 IEEE MTT-S International Microwave Symposium Digest (MTT)*, pp. 1–3, June 2014.

[MAY 08] MAYER L.W., SCHOLTZ A.L., "A dual-band HF/UHF antenna for RFID tags", *IEEE 68th Vehicular Technology Conference, 2008. VTC 2008-Fall*, pp. 1–5, September 2008.

[MAY 09a] MAYER L.W., SCHOLTZ A.L., "Circularly polarized patch antenna with high Tx/Rx-separation", *IEEE RFID 2009, 2009 IEEE International Conference on RFID*, Orlando, USA, pp. 213–216, April 2009.

[MAY 09b] MAYER L.W., LANGWIESER R., SCHOLTZ A.L., "Evaluation of passive carrier-suppression techniques for UHF RFID systems", *2009 IEEE MTT-S International Microwave Workshop on Wireless Sensing, Local Positioning and RFID*, Cavtat, Croatia, September 2009.

[MAY 12] MAYER L.W., Dual-band antenna, US Patent 8,174,454, May 8, 2012.

[MCS 92] MCSPADDEN J.O., YOO T., CHANG K., "Theoretical and experimental investigation of a rectenna element for microwave power transmission", *IEEE Transactions on Microwave Theory and Techniques*, vol. 40, no. 12, pp. 2359–2366, 1992.

[MOB 10] MOBASHSHER A.T., ISLAM M.T., MISRAN N., "A novel high-gain dual-band antenna for RFID reader applications", *IEEE Antennas and Wireless Propagation Letters*, vol. 9, pp. 653–656, 2010.

[MOH 12] MOHAMED B., FERGANI B., TEDJINI S. *et al.*, "Simulation and measurements of physical layer parameters for passive UHF RFID tags toward collision detection and canceling", *2012 IEEE International Conference on RFID-Technologies and Applications (RFID-TA)*, pp. 383–388, November 2012.

[NAW 12] NAWALE S.D., SARAWADE N.P., "RFID vapor sensor: beyond identification", *2012 Sixth International Conference on Sensing Technology (ICST)*, pp. 248–253, December 2012.

[NAZ 12] NAZARI SHIREHJINI A.A., YASSINE A., SHIRMOHAMMADI S., "An RFID-based position and orientation measurement system for mobile objects in intelligent environments", *IEEE Transactions on Instrumentation and Measurement*, vol. 61, no. 6, pp. 1664–1675, 2012.

[NGU 12] NGUYEN D.S., LE X.C., PHAM T.T. *et al.*, "Novel design of RFID UHF passive tag for wideband applications by direct and contactless chip connection", *IEEE 2012 International Conference on RFID – Technologies and Applications (RFID – TA) (IEEE RFID-TA 2012)*, Nice, France, France, pp. 131–136, November 2012.

[NGU 13] NGUYEN S.D., PHAM T.T., BLANC F. *et al.*, "Approach for quality detection of food by RFID-based wireless sensor tag", *Electronics Letters*, vol. 49, no. 25, pp. 1588–1589, 5 December 2013.

[NIK 06] NIKITIN P.V., RAO K.V.S., "Theory and measurement of backscattering from RFID tags", *IEEE Antennas and Propagation Magazine*, vol. 48, no. 6, pp. 212–218, December 2006.

[NIK 07a] NIKNEJAD A.M., *Electromagnetics for High-Speed Analog and Digital Communication Circuits*, Cambridge University Press, Cambridge, 2007.

[NIK 07b] NIKITIN P.V., RAO K.V.S., MARTINEZ R.D., "Differential RCS of RFID tag", *Electronics Letters*, vol. 43, no. 8, pp. 431–432, 2007.

[NIK 09a] NIKITIN P.V., RAO K., "Effect of Gen2 protocol parameters on RFID tag performance", *2009 IEEE International Conference on RFID*, pp. 117–122, April 2009.

[NIK 09b] NIKITIN P.V., RAO K.V.S., "Harmonic scattering from passive UHF RFID tags", *IEEE Antennas and Propagation Society International Symposium, 2009. APSURSI'09*, pp. 1–4, June 2009.

[NIK 09c] NIKITIN P.V., RAO K.V.S., MARTINEZ R. *et al.*, "Sensitivity and impedance measurements of UHF RFID chips", *IEEE Transactions on Microwave Theory and Techniques*, vol. 57, no. 5, pp. 1297–1302, 2009.

[NIK 12a] NIKITIN P.V., "Leon Theremin (Lev Termen)", *IEEE Antennas and Propagation Magazine*, vol. 54, no. 5, pp. 252–257, 2012.

[NIK 12b] NIKITIN P.V., RAO K.VS., LAM S., "UHF RFID tag characterization: overview and state-of-the-art", *Annual Symposium of the Antenna Measurement Techniques Association AMTA*, pp. 1–6, 2012.

[NIK 12c] NIKITIN P.V., RAMAMURTHY S., MARTINEZ R. *et al.*, "Passive tag-to-tag communication", *2012 IEEE International Conference on RFID (RFID)*, pp. 177–184, 2012.

[NIK 17] NIKITIN P., "Self-reconfigurable RFID reader antenna", *2017 IEEE International Conference on RFID (RFID)*, Phoenix, AZ, 2017, pp. 88–95, 2017. doi: 10.1109/RFID.2017.7945592

[NTI 14] NTIA-USA, "National Telecommunications and Information Administration 2700–2900 MHz", available at: http://www.ntia.doc.gov/files/ntia/publications/compendium/2700.00-2900.00_01MAR14.pdf, 2014.

[NXP 13] NXP, "UCODE G2XM and G2XL Product data sheet", available at: http://www.nxp.com/documents/data\s\do3(s)heet/SL3ICS1002\s\do3(1)202.pdf, 2013.

[ORE 11] ORECCHINI G., ALIMENTI F., PALAZZARI V. *et al.*, "Design and fabrication of ultra-low cost radio frequency identification antennas and tags exploiting paper substrates and inkjet printing technology", *IET Microwaves, Antennas Propagation*, vol. 5, no. 8, pp. 993–1001, 2011.

[OSE 11] OSEO, "Investissements d'avenir", available at: http://www.investissementsdavenir-oseo.fr/, 2011.

[PAR 04] PARK J.-Y., HAN S.-M., ITOH T., "A rectenna design with harmonic-rejecting circular-sector antenna", *IEEE Antennas and Wireless Propagation Letters*, vol. 3, no. 1, pp. 52–54, 2004.

[PAR 10a] PARET D., *RFID at Ultra and Super High Frequencies: Theory and Application*, Wiley Publishing, 2010.

[PAR 10b] PAREDES F., GONZALEZ G.Z., BONACHE J. *et al.*, "Dual-band impedance-matching networks based on split-ring resonators for applications in RF identification (RFID)", *IEEE Transactions on Microwave Theory and Techniques*, vol. 58, no. 5, pp. 1159–1166, 2010.

[PER 12] PERRET E., TEDJINI S., NAIR R.S., "Design of antennas for UHF RFID tags", *Proceedings of the IEEE*, vol. 100, no. 7, pp. 2330–2340, 2012.

[PHO 10] PHONGCHAROENPANICH C., SUWALAK R., "Dual-band RFID-reader antenna using annular plate with curved and rectangular slots", *2010 International Conference on Electromagnetics in Advanced Applications (ICEAA)*, pp. 633–636, September 2010.

[POP 13] POPOVIC Z., FALKENSTEIN E., ZANE R., "Low-power density wireless powering for battery-less sensors", *Radio and Wireless Symposium (RWS), 2013 IEEE*, pp. 31–33, January 2013.

[POZ 05] POZAR D.M., *Microwave Engineering*, 3rd ed., Wiley, 2005.

[PRE 08] PRERADOVIC S., BALBIN I., KARMAKAR N.C. *et al.*, "Chipless frequency signature based RFID transponders", *Microwave Conference, 2008. EuMC 2008. 38th European*, pp. 1723–1726, October 2008.

[PRE 09] PRERADOVIC S., BALBIN I., KARMAKAR N.C. *et al.*, "Multiresonator-based Chipless RFID system for low-cost item tracking", *IEEE Transactions on Microwave Theory and Techniques*, vol. 57, no. 5, pp. 1411–1419, 2009.

[PSY 08] PSYCHOUDAKIS D., MOULDER W., CHEN C.-C. *et al.*, "A portable low-power harmonic radar system and conformal tag for insect tracking", *IEEE Antennas and Wireless Propagation Letters*, vol. 7, pp. 444–447, 2008.

[RAI 17] RAIN RFID, "Welcome to RAIN RFID", available at: http://rainrfid.org/-RAIN RFID, 2017.

[RAM 13] RAMOS A., LAZARO A., GIRBAU D., "Semi-passive time-domain UWB RFID system", *IEEE Transactions on Microwave Theory and Techniques*, vol. 61, no. 4, pp. 1700–1708, 2013.

[RAO 05] RAO K.V.S., NIKITIN P.V., LAM S.F., "Antenna design for UHF RFID tags: a review and a practical application", *IEEE Transactions on Antennas and Propagation*, vol. 53, no. 12, pp. 3870–3876, 2005.

[RAP 13] RAPIDTRAK UHF TAG, Close proximity, Omni-directional reading of hanging items at retail points-of-sale (POS) TAGSYS RFID, King of Prussia, PA, USA, 2013.

[RAS 14a] RASILAINEN K., ILVONEN J., LEHTOVUORI A. *et al.*, "On design and evaluation of harmonic transponders", *IEEE Transactions on Antennas and Propagation*, vol. 63, no. 1, pp. 15–23, January 2014, doi: 10.1109/TAP.2014.2366193.

[RAS 14b] RASILAINEN K., ILVONEN J., VIIKARI V., "Antenna matching at harmonic frequencies to complex load impedance", *IEEE Antennas and Wireless Propagation Letters*, vol. 14, pp. 535–538, 2014, doi: 10.1109/LAWP.2014.2370760.

[RAZ 98] RAZAVI B., *RF Microelectronics*, vol. 1, Prentice Hall, New Jersey, 1998.

[RFI 17] RFID XPLORER – CISC SEMICONDUCTOR, "CISC Semiconductor GmbH", available at: https://www.cisc.at/products/rfid-xplorer/, 2017.

[ROB 12] ROBERG M., REVEYRAND T., RAMOS I. *et al.*, "High-efficiency harmonically terminated diode and transistor rectifiers", *IEEE Transactions on Microwave Theory and Techniques*, vol. 60, no. 12, pp. 4043–4052, 2012.

[ROM 09] ROMERO H.P., REMLEY K.A., WILLIAMS D.F. *et al.*, "Electromagnetic measurements for counterfeit detection of radio frequency identification cards", *IEEE Transactions on Microwave Theory and Techniques*, vol. 57, no. 5, pp. 1383–1387, 2009.

[ROS 12] ROSELLI L., ALIMENTI F., VIRILI M. *et al.*, "Feasibility study of a fully organic frequency doubler for harmonic RFID applications", *2012 IEEE 12th Topical Meeting on Silicon Monolithic Integrated Circuits in RF Systems (SiRF)*, pp. 203–206, 2012.

[SAT 09] SATO, AD-222 RFID Inlays, available at: http://www.satosolutions.com/Images/Users/Products/RFID/LabelSpec/GL_AveryAD222_Ver1.1.pdf, 2009.

[SCH 14] SCHEELER R., KORHUMMEL S., POPOVIC Z., "A dual-frequency ultralow-power efficient 0.5-g rectenna", *IEEE Microwave Magazine*, vol. 15, no. 1, pp. 109–114, 2014.

[SHE 75] SHEFER J., Harmonic radar detecting and ranging system for automotive vehicles, US Patent RE28, 302, available at: http://www.google.com.ar/patents/USRE28302, January 14, 1975.

[SID 17] SIDEBOTTOM A., THORNTON A., TOMPSON L. *et al.*, "A systematic review of tagging as a method to reduce theft in retail environments", *Crime Science An Interdisciplinary Journal*, vol. 6, no. 7, available at: https://doi.org/10.1186/s40163-017-0068-y, 2017.

[SKA 09] SKALI S., CHANTEPY C., TEDJINI S., "On the measurement of the delta Radar Cross Section (ΔRCS) for UHF tags", *2009 IEEE International Conference on RFID*, pp. 346–351, April 2009.

[SON 14] SONG J., SALMI J., VIIKARI V. *et al.*, "Maximum likelihood estimation for passive wireless intermodulation communication sensors", *IEEE Sensors Journal*, vol. 15, no. 4, pp. 2280–2286, April 2014, doi: 10.1109/JSEN.2014.2372058.

[SPI 12] SPINNAKER-PROJECT, "Item-level sensing, anywhere, anytime", available at: http://www.spinnaker-rfid.com/, 2012.

[STA 73] STARAS H., SHEFER J., Harmonic radar detecting and ranging system for automotive vehicles, US Patent 3,781,879, 2 December 1973.

[SUN 13] SUN H., GUO Y.-X., HE M. *et al.*, "A dual-band rectenna using broadband Yagi antenna array for ambient RF power harvesting", *IEEE Antennas and Wireless Propagation Letters*, vol. 12, pp. 918–921, 2013.

[TAG 15] TAGSYS RFID, "Radio Frequency Identification – Tagsys RFID", available at: www.tagsysrfid-fits.com, 2015.

[TAG 17] TAGFORMANCE-PRO, "Voyantic Ltd", available at: http://voyantic.com/products/tagformance-pro

[TEC 16] TECHNOLOGIES A., "Agilent Command Expert", available at: literature.cdn.keyword.com, 2016.

[TED 13] TEDJINI S., KARMAKAR N., PERRET E. *et al.*, "Hold the chips: chipless technology, an alternative technique for RFID", *IEEE Microwave Magazine*, vol. 14, no. 5, pp. 56–65, 2013.

[TED 16] TEDJINI S., ANDIA-VERA G., ZURITA M. *et al.*, "Augmented RFID Tags", *2016 IEEE Topical Conference on Wireless Sensors and Sensor Networks (WiSNet)*, Austin, TX, pp. 67–70, 2016, doi: 10.1109/WISNET.2016.7444324.

[TRO 09] TROTTER M.S., GRIFFIN J.D., DURGIN G.D., "Power-optimized waveforms for improving the range and reliability of RFID systems", *2009 IEEE International Conference on RFID*, pp. 80–87, April 2009.

[TUM 89] TUMMALA R.R., RYMASZEWSKI E.J., *Microelectronics Packaging Handbook*, 1989, doi: https://doi.org/10.1007/978-1-4615-4086-1_1.

[UCO 17] UCODE, "DNA NXP", available at: https://www.nxp.com/products/identification-and-security/smart-label-and-tag-ics/ucode-dna:MC_1436185909808-NXP, 2017.

[UPM 13a] UPM RAFSEC MEMORYSTICK, "Wet inlay, Global UHF C1G2 EPC, UPM", Dixon, available at: http://www.rf-id.ru/img/content/3000917.pdf, 2013.

[UPM 13b] UPM RAFLATEC DOGBONE, "General purpose high-end product for global supply chain management and item-level use", UPM RAFLATAC, Dixon, available at: http://www.upmraflatac.com/europe/eng/images/51\s\do3(6)8732.pdf, 2013.

[UPM 13c] UPM RAFLATAC FROG, "Orientation insensitive product for global supply chain. Suited for corrugated materials and plastics", UPM RAFLATAC, Dixon, available at: http://www.upmraflatac.com/europe/eng/images/51\s\do3(6)8734.pdf, 2013.

[UPM 13d] UPM RAFSEC FROG, available at: http://www.webermarking.com/images/Dryinlay, Dixon, 2013.

[USA 04] USAMI M., "An ultra small RFID chip: mu-chip", *Digest of Papers. 2004 IEEE Radio Frequency Integrated Circuits (RFIC) Symposium, 2004*, pp. 241–244, June 2004.

[VAZ 10] VAZ A., UBARRETXENA A., ZALBIDE I. *et al.*, "Full passive UHF tag with a temperature sensor suitable for human body temperature monitoring", *IEEE Transactions on Circuits and Systems II: Express Briefs*, vol. 57, no. 2, pp. 95–99, February 2010.

[VIR 14] VIRILI M., CASULA G., MARIOTTI C., "7.5–15 MHz organic frequency doubler made with pentacene-based diode and paper substrate", *2014 IEEE MTT-S International Microwave Symposium (IMS)*, pp. 1–4, June 2014.

[WAN 07] WANG W., LOU S., CHUI K. *et al.*, "Single-chip UHF RFID reader in 0.18- um CMOS", *IEEE Custom Integrated Circuits Conference, 2007. CICC'07*, pp. 111–114, 2007.

[WIR 17] WIRELESS TEMPERATURE SENSORS, "RF Micron", available at: http://rfmicron.com/temp-sensor/#, 2017.

[YAN 08] YAN L., ZHANG Y., YANG L. *et al.*, "The Internet of things: from RFID to the next-generation pervasive networked systems", *Wireless Networks and Mobile Communications*, CRC Press, Boca Raton, 2008.

[YU 10] YU F., LYON K.G., KAN E.C., "A novel passive RFID transponder using harmonic generation of non-linear transmission lines", *IEEE Transactions on Microwave Theory and Techniques*, vol. 58, no. 12, pp. 4121–4127, 2010.

[ZHA 10] ZHANG M., YAN S., "A novel dual-band dipole antenna for RFID tag", *2010 International Conference on Logistics Engineering and Intelligent Transportation Systems (LEITS)*, pp. 1–4, November 2010.

Index

U, V

Other titles from

iSTE

in

Networks and Telecommunications

2017

BENSLAMA Malek, BENSLAMA Achour, ARIS Skander
Quantum Communications in New Telecommunications Systems

HILT Benoit, BERBINEAU Marion, VINEL Alexey, PIROVANO Alain
Networking Simulation for Intelligent Transportation Systems: High Mobile Wireless Nodes

LESAS Anne-Marie, MIRANDA Serge
The Art and Science of NFC Programming
(Intellectual Technologies Set – Volume 3)

2016

AL AGHA Khaldoun, PUJOLLE Guy, ALI-YAHIYA Tara
Mobile and Wireless Networks
(Advanced Network Set – Volume 2)

BATTU Daniel
Communication Networks Economy

BENSLAMA Malek, BATATIA Hadj, MESSAI Abderraouf
Transitions from Digital Communications to Quantum Communications: Concepts and Prospects

CHIASSERINI Carla Fabiana, GRIBAUDO Marco, MANINI Daniele
Analytical Modeling of Wireless Communication Systems (Stochastic Models in Computer Science and Telecommunication Networks Set – Volume 1)

EL FALLAH SEGHROUCHNI Amal, ISHIKAWA Fuyuki, HÉRAULT Laurent, TOKUDA Hideyuki
Enablers for Smart Cities

PEREZ André
VoLTE and ViLTE

2015

BENSLAMA Malek, KIAMOUCHE Wassila, BATATIA Hadj
Connections Management Strategies in Satellite Cellular Networks

BENSLAMA Malek, BATATIA Hadj, BOUCENNA Mohamed Lamine
Ad Hoc Networks Telecommunications and Game Theory

BERTHOU Pascal, BAUDOIN Cédric, GAYRAUD Thierry, GINESTE Matthieu
Satellite and Terrestrial Hybrid Networks

CUADRA-SANCHEZ Antonio, ARACIL Javier
Traffic Anomaly Detection

LE RUYET Didier, PISCHELLA Mylène
Digital Communications 1: Source and Channel Coding

PEREZ André
LTE and LTE Advanced: 4G Network Radio Interface

PISCHELLA Mylène, LE RUYET Didier
Digital Communications 2: Digital Modulations

PUJOLLE Guy
Software Networks (Advanced Network Set – Volume 1)

2014

ANJUM Bushra, PERROS Harry
Bandwidth Allocation for Video under Quality of Service Constraints

BATTU Daniel
New Telecom Networks: Enterprises and Security

BEN MAHMOUD Mohamed Slim, GUERBER Christophe, LARRIEU Nicolas, PIROVANO Alain, RADZIK José
Aeronautical Air–Ground Data Link Communications

BITAM Salim, MELLOUK Abdelhamid
Bio-inspired Routing Protocols for Vehicular Ad-Hoc Networks

CAMPISTA Miguel Elias Mitre, RUBINSTEIN Marcelo Gonçalves
Advanced Routing Protocols for Wireless Networks

CHETTO Maryline
Real-time Systems Scheduling 1: Fundamentals
Real-time Systems Scheduling 2: Focuses

EXPOSITO Ernesto, DIOP Codé
Smart SOA Platforms in Cloud Computing Architectures

MELLOUK Abdelhamid, CUADRA-SANCHEZ Antonio
Quality of Experience Engineering for Customer Added Value Services

OTEAFY Sharief M.A., HASSANEIN Hossam S.
Dynamic Wireless Sensor Networks

PEREZ André
Network Security

PERRET Etienne
Radio Frequency Identification and Sensors: From RFID to Chipless RFID

REMY Jean-Gabriel, LETAMENDIA Charlotte
LTE Standards
LTE Services

TANWIR Savera, PERROS Harry
VBR Video Traffic Models

VAN METER Rodney
Quantum Networking

XIONG Kaiqi
Resource Optimization and Security for Cloud Services

2013

ASSING Dominique, CALÉ Stéphane
Mobile Access Safety: Beyond BYOD

BEN MAHMOUD Mohamed Slim, LARRIEU Nicolas, PIROVANO Alain
Risk Propagation Assessment for Network Security: Application to Airport Communication Network Design

BEYLOT André-Luc, LABIOD Houda
Vehicular Networks: Models and Algorithms

BRITO Gabriel M., VELLOSO Pedro Braconnot, MORAES Igor M.
Information-Centric Networks: A New Paradigm for the Internet

BERTIN Emmanuel, CRESPI Noël
Architecture and Governance for Communication Services

DEUFF Dominique, COSQUER Mathilde
User-Centered Agile Method

DUARTE Otto Carlos, PUJOLLE Guy
Virtual Networks: Pluralistic Approach for the Next Generation of Internet

FOWLER Scott A., MELLOUK Abdelhamid, YAMADA Naomi
LTE-Advanced DRX Mechanism for Power Saving

JOBERT Sébastien *et al.*
Synchronous Ethernet and IEEE 1588 in Telecoms: Next Generation Synchronization Networks

MELLOUK Abdelhamid, HOCEINI Said, TRAN Hai Anh
Quality-of-Experience for Multimedia: Application to Content Delivery Network Architecture

NAIT-SIDI-MOH Ahmed, BAKHOUYA Mohamed, GABER Jaafar, WACK Maxime
Geopositioning and Mobility

PEREZ André
Voice over LTE: EPS and IMS Networks

2012

AL AGHA Khaldoun
Network Coding

BOUCHET Olivier
Wireless Optical Communications

DECREUSEFOND Laurent, MOYAL Pascal
Stochastic Modeling and Analysis of Telecoms Networks

DUFOUR Jean-Yves
Intelligent Video Surveillance Systems

EXPOSITO Ernesto
Advanced Transport Protocols: Designing the Next Generation

JUMIRA Oswald, ZEADALLY Sherali
Energy Efficiency in Wireless Networks

KRIEF Francine
Green Networking

PEREZ André
Mobile Networks Architecture

2011

BONALD Thomas, FEUILLET Mathieu
Network Performance Analysis

CARBOU Romain, DIAZ Michel, EXPOSITO Ernesto, ROMAN Rodrigo
Digital Home Networking

CHABANNE Hervé, URIEN Pascal, SUSINI Jean-Ferdinand
RFID and the Internet of Things

GARDUNO David, DIAZ Michel
Communicating Systems with UML 2: Modeling and Analysis of Network Protocols

LAHEURTE Jean-Marc
Compact Antennas for Wireless Communications and Terminals: Theory and Design

RÉMY Jean-Gabriel, LETAMENDIA Charlotte
Home Area Networks and IPTV

PALICOT Jacques
Radio Engineering: From Software Radio to Cognitive Radio

PEREZ André
IP, Ethernet and MPLS Networks: Resource and Fault Management

TOUTAIN Laurent, MINABURO Ana
Local Networks and the Internet: From Protocols to Interconnection

2010

CHAOUCHI Hakima
The Internet of Things

FRIKHA Mounir
Ad Hoc Networks: Routing, QoS and Optimization

KRIEF Francine
Communicating Embedded Systems / Network Applications

2009

CHAOUCHI Hakima, MAKNAVICIUS Maryline
Wireless and Mobile Network Security

VIVIER Emmanuelle
Radio Resources Management in WiMAX

2008

CHADUC Jean-Marc, POGOREL Gérard
The Radio Spectrum

GAÏTI Dominique
Autonomic Networks

LABIOD Houda
Wireless Ad Hoc and Sensor Networks

LECOY Pierre
Fiber-optic Communications

MELLOUK Abdelhamid
End-to-End Quality of Service Engineering in Next Generation Heterogeneous Networks

PAGANI Pascal *et al.*
Ultra-wideband Radio Propagation Channel

2007

BENSLIMANE Abderrahim
Multimedia Multicast on the Internet

PUJOLLE Guy
Management, Control and Evolution of IP Networks

SANCHEZ Javier, THIOUNE Mamadou
UMTS

VIVIER Guillaume
Reconfigurable Mobile Radio Systems

Printed and bound by CPI Group (UK) Ltd, Croydon, CR0 4YY

15/09/2026

14962345-0001